Industrial Processing
with Membranes

INDUSTRIAL PROCESSING WITH MEMBRANES

EDITED BY

Robert E. Lacey

Southern Research Institute
Birmingham, Alabama

and

Sidney Loeb

Negev Institute for Arid Zone Research
Beersheva, Israel

WILEY-INTERSCIENCE, a Division of John Wiley & Sons, Inc.
New York · London · Sydney · Toronto

CHEMISTRY

Library of Congress Catalog Card Number: 79-178145

ISBN 0-471-51136-6

Printed in the United States of America.

10 9 8 7 6 5 4 3 2 1

To my late wife Peg for her
patience and encouragement
with the visions and revisions
that went into this book.

REL

Preface

The use of membrane processes to achieve chosen purposes should come naturally to man, since life depends on transfers of materials through many kinds of membranes in the body. However, the development of membrane processes to the point at which they can be used for industrial applications is relatively recent. The initial impetus to the development of membrane processes stemmed largely from the research on such processes for desalination of salt water, much of which was supported and encouraged by the Office of Saline Water of the United States Department of the Interior. After the development of processes such as electrodialysis and reverse osmosis for desalination of salt water, these processes have begun to be used for industrial separations and it seems certain that such processes will find wide application in the chemical process industries, in the food industry, and for the treatment of wastes.

One purpose of this book is to provide the reader with information that will aid in determining whether membrane processes should be considered for a given industrial separation and, if so, which type of membrane process is most promising. Another aim of the book is to present the basic principles governing membrane processing to aid in the design and operation of membrane processing plants (or pilot plants). Limitations of space prohibit as comprehensive a treatment of the topics in the book as many of the authors would have desired, but extensive bibliographies are given to aid the reader in pursuing particular subjects in detail. Also because of lack of space, some membrane processes could not be discussed in this book. We chose to omit discussions of dialysis and pervaporation because these processes have been treated in detail by Tuwiner in *Diffusion and Membrane Technology* (Reinhold, New York) and by C. Y. Choo in Volume VI of *Advances in Petroleum Chemistry and Refining* (Wiley-Interscience, New York).

The book is divided into two sections. Part I discusses electrically driven processes, and Part II, pressure-driven processes. Chapter 1 is an introduction to electromembrane processing in which its development is discussed and various types of electromembrane processes are described briefly. Chapter 2 presents information about the ion-exchange membranes used in electromembrane processing and about the basic physicochemical and hydrodynamic principles needed to understand the processes. Chapter 3 discusses economic and engineering aspects of importance to electromembrane processing. These chapters were planned to provide the reader with information about the basic principles of electromembrane processing so the illustrative examples of industrial applications in the next three chapters can be easily understood. Chapters 4, 5, and 6 describe applications of electrically driven processes to the demineralization of whey, the treatment of certain waste streams in the pulp and paper industries, and the concentration of electrolytes for industrial purposes.

Part II on pressure-driven processes is organized similarly. Chapter 7 is an introductory chapter by Professor Reid, whose pioneering work introduced the world to the concept of reverse osmosis. Chapter 8 discusses the membranes used in ultrafiltration and reverse osmosis and presents the basic principles necessary to understand the processes. Chapter 9 discusses the cost considerations in processing by ultrafiltration and reverse osmosis. The next three chapters are illustrative examples of the application of pressure-driven processes to industrial problems. Chapter 10 and 11 discuss the use of reverse osmosis in the food industry and in the pulp and paper industry, and Chapter 12 discusses the use of ultrafiltration and reverse osmosis for the treatment of industrial wastes. Finally, Chapter 13 is a comprehensive treatment of the pressure-driven process of gas permeation.

We regret that some types of membrane processes (e.g., dialysis, pervaporation, and Donnan dialysis) could not be treated in this book, and that many interesting applications of membrane processes, such as medical and pharmaceutical applications, could not be described. Despite the limitations in content, we hope the book serves a useful purpose in that a number of scientists experienced in membrane technology are able to share their ideas with others not yet as experienced.

We give thanks to the authors for their hard work in preparing their chapters. We also want to thank the management of Southern Research Institute for their cooperation and encouragement. We want to give special thanks to Dr. Sivert N. Glarum for his aid in some of the editorial tasks, and to the persevering secretarial staff at the Institute, whose patience we must have tried sorely.

Robert E. Lacey
Sidney Loeb

Birmingham, Alabama
Beersheva, Israel
July 1971

Contents

ix

Industrial Processing
with Membranes

Part 1   Electrically Driven Membrane Processes

Chapter I Basis of Electro-membrane Processes

Robert E. Lacey*

Electrodialysis is the transport of ions through membranes as a result of an electrical driving force. It is the phenomenon underlying all electromembrane processes. If nonselective membranes that are permeable to ions (e.g., cellophane) are used in electrodialytic processes, electrolytes can be separated from nonelectrolytes. If the membranes are more permeable to cations than to anions or vice versa (e.g., ion-exchange membranes), the concentrations of ionic species in solutions can be increased or decreased by electrodialysis, so that practical concentration or depletion of electrolyte solutions is possible. With membranes of relatively recent origin that are more permeable to univalent ions than to multivalent ions, electrodialysis can be used to simultaneously separate and concentrate univalent ions from solutions containing mixtures of uni- and multivalent electrolytes.

This chapter presents a brief introduction to the basic principles common to all electrically driven membrane processes, a discussion of some of the variants of electrodialysis, and the future possibilities of membrane processes. These topics are discussed in detail later in this book and only briefly here.

I. HISTORICAL DEVELOPMENT OF ELECTRODIALYSIS

Although simple electrodialysis processes with nonselective membranes were studied in Germany during the early part of this century and the processes were reviewed and described in 1931 by Prausnitz and Reitstotter,[1] it

* Southern Research Institute, Birmingham, Alabama.

was not until 1939 that Manegold and Kalauch[2] pointed out the advantages of the use of highly selective membranes to achieve demineralization by electrodialysis. In 1940, Meyer and Strauss[3] first proposed an electro-dialysis process in which a number of anion-selective and cation-selective membranes were arranged between a pair of electrodes so that anion-selective membranes alternated with cation-selective membranes to form many parallel solution compartments, as illustrated in Figure 1. With such

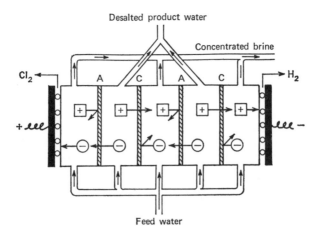

Figure 1.   The electrodialysis process for desalination. A, anion-permeable membrane; C, cation-permeable membrane.

a multicompartment electrodialyzer, demineralization (or concentration) of solutions could be achieved (in a manner described later) in many compart-ments with only one pair of electrodes. With multicompartment electro-dialysis, the irreversibilities represented by the decomposition potentials at the electrodes could be distributed over many demineralizing compartments and thus minimized, and the problems of handling the products formed at the electrodes could be minimized. With the development in the late 1940's of stable ion-exchange membranes of low electrical resistance, multi-compartment electrodialysis became a practical process for demineralizing and concentrating solutions of electrolytes. During the 1950's the United States Department of the Interior's Office of Saline Water directly supported much research and development of the electrodialysis process for de-mineralizing saline water and encouraged research on the process by private companies. During the same span of years, the Netherlands National Research Organization, TNO, and the South African Council for Scientific and Industrial Research developed the electrodialysis process for demineralizing saline waters originating in mines. In the years following 1955, the Office of European Economic Cooperation also supported research and development on the process, and in the 1960's the Institute for Arid Zone Research at Beersheva, Israel, made further contributions to the development of electrodialysis. In the late 1950's and the 1960's, several Japanese manufacturers developed the electrodialysis process as a means of concentrating seawater for use as a brine for the chlor-alkali industries.

A book on electrodialysis, edited by Wilson[4] and published in 1960, describes the process primarily in terms of its use in saline water conversion. Chapters in books by Tuwiner[5] and Spiegler[6] also describe electrodialysis mainly as a desalination process. Reviews by Lakshiminarayaniah[7] and by Friedlander and Rickles[8] discuss the preparation of ion-selective membranes and the phenomena involved in membrane transport. Chapters in a book by Helferrich[9] and in a privately published book[10] treat in detail the physical chemistry of membrane transport.

II. ION-EXCHANGE MEMBRANES

Changes in the concentration of electrolytes in a solution can be accomplished by electrodialysis, if ion-exchange membranes are used. Ion-exchange membranes are ion exchangers in film form. There are two types: anion-exchange and cation-exchange membranes. Anion-exchange membranes contain cationic groups fixed to the resin matrix. The fixed cations are in electroneutrality with mobile anions in the interstices of the resin. When such a membrane is immersed in a solution of an electrolyte, the anions in solution can intrude into the resin matrix and replace the anions initially present, but the cations are prevented from entering the matrix by the repulsion of the cations affixed to the resin.

Cation-exchange membranes are similar. They contain fixed anionic groups that permit intrusion and exchange of cations from an external source, but exclude anions. This type of exclusion is usually called Donnan exclusion in honor of the pioneering work of F. G. Donnan. It is discussed in detail in Chapter 2.

The details of methods for making ion-exchange membranes have been reviewed recently.[4,7,8] In addition, reviews of methods of manufacturing ion-exchange membranes are included in the *Annual Reviews of Ion Exchange Materials* published by *Industrial Engineering Chemistry* magazine. Heterogeneous membranes have been made by incorporating ion-exchange particles into film-forming resins (*a*) by dry molding or calendering mixtures of the ion-exchange and film-forming materials; (*b*) by dispersing the ion-exchange

Table 1. Reported Properties of Ion-Exchange Membranes*

Manufacturer and Designation	Type of Membrane	Area Resistance (ohm-cm^2)	Transference Number of Counterion[a]	Strength	Approximate Thickness (mils)	Dimensional Changes on Wetting and Drying (%)	Size Available
AMF[b]		(0.6 N KCl)		Mullen burst (psi)			
C-60	Cat-exch	5 ± 2	0.80 (0.5/1.0 N KCl)	45	12	10–13	44-in. wide rolls
C-100	Cat-exch	7 ± 2	0.90 (0.5/1.0 N KCl)	60	8.5		
A-60	An-exch	6 ± 2	0.80 (0.5/1.0 N KCl)	45	12	12–15	44-in. wide rolls
A-100	An-exch	8 ± 2	0.90 (0.5/1.0 N KCl)	55	9		
ACI[c]		(0.5 N NaCl)		Tensile strength (kg/mm^2)			
CK-1	Cat-exch	1.4	0.85 (0.25/0.5 N NaCl)	2 to 2.4	9	15–23	44 × 44 in.
DK-1	Cat-exch	1.8	0.85 (0.25/0.5 N NaCl)		9		
CA-1	An-exch	2.1	0.92 (0.25/0.5 N NaCl)	2 to 2.3	9	12–18	44 × 44 in.
DA-1	An-exch	3.5	0.92 (0.25/0.5 N NaCl)		9		
AGC[d]		(0.5 N NaCl)		Mullen burst (psi)			
CMV	Cat-exch	3	0.93 (0.5/1.0 N NaCl)	ca 180	6		
CSV	Cat-exch	10	0.92 (0.5/1.0 N NaCl)	ca 180	12	<2	44-in. wide rolls
AMV	An-exch	4	0.95 (0.5/1.0 N NaCl)	ca 150	6		
ASV	An-exch	5	0.95 (0.5/1.0 N NaCl)	ca 150	6		
IC[f]		(0.1 N NaCl)		Mullen burst (psi)			
MC-3142	Cat-exch	12	0.94 (0./51.0 N NaCl)	ca 200	8		
MC-3235	Cat-exch	18	0.95 (0./10.2 N NaCl)	ca 165	12	<3[g]	40 × 120 in.
MC-3470	Cat-exch	35	0.98 (0./10.2 N NaCl)	ca 200	8		

		(0.1 N NaCl) / (0.5 N NaCl)		Mullen burst (psi)			
MA-3148	An-exch	20	0.90 (0.5/1.0 N NaCl)	ca 200	8		
MA-3236	An-exch	120	0.93 (0.5/1.0 N NaCl)	ca 165	12	<3[g]	40 × 120 in.
IM-12	An-exch[h]	12	0.96 (0.1/0.2 N NaCl)[g]	ca 145	6[g]	Not given	
MA-3475R	An-exch	11	0.99 (0.5/1.0 N NaCl)	ca 200	14	Not given	
II[i]		(0.1 N NaCl)		Mullen burst (psi)			
CR-61	Cat-exch	11	0.93 (0.1/0.2 N NaCl)[b]	115	23	Cracks on drying	18 × 40 in.
AR-111A	An-exch	11	0.93 (0.1/0.2 N NaCl)[b] (by electrophoretic method in 0.5 N NaCl)	125	24		
TSC[j]		(0.5 N NaCl)		Mullen burst (psi)			
CL-2.5T	Cat-exch	3	0.98	ca 80	6	Not given	40 × 50 in.
CLS-25T	Cat-exch[k]	3	0.98	ca 80	6	Not given	40 × 50 in.
AV-4T	An-exch	4	0.98	ca 150	7	Not given	40 × 50 in.
AVS-4T	An-exch[k]	5	0.98	ca 140	7	Not given	40 × 50 in.

* Properties are those reported by manufacturer, except for those membranes designated with footnote g.
[a] Calculated from concentration potentials measured between solutions of the two normalities listed.
[b] American Machine and Foundry Co., Stamford, Connecticut.
[c] Asahi Chemical Industry, Ltd., Tokyo, Japan.
[d] Asahi Glass Co., Ltd., Tokyo, Japan.
[e] Membranes that are selective for univalent (over multivalent) ions.
[f] Ionac Chemical Co., Birmingham, New Jersey.
[g] Measured at Southern Research Institute.
[h] Special anion-exchange membrane that is highly diffusive to acids.
[i] Ionics, Inc., Cambridge, Massachusetts.
[j] Tokuyama Soda Co., Ltd., Tokyo, Japan.
[k] Univalent selective membranes.

7

material in a solution of the film-forming polymer, then casting films from the solution and evaporating the solvent; and (c) by dispersing the ion-exchange material in a partially polymerized film-forming polymer, casting films, and completing the polymerization.

Heterogeneous membranes with usefully low electrical resistances contain more than 65% by weight of the crosslinked ion-exchange particles. Since these ion-exchange particles swell when immersed in water, it has been difficult to achieve adequate mechanical strength and freedom from distortion combined with low electrical resistance.

To overcome these and other difficulties with heterogeneous membranes, homogeneous membranes were developed in which the ion-exchange component forms a continuous phase throughout the resin matrix. The general methods of preparing homogeneous membranes are as follows:

- Polymerization of mixtures of reactants (e.g., phenol, phenolsulfonic acid, and formaldehyde) that can undergo condensation polymerization. At least one of the reactants must contain a moiety that either is, or can be made, anionic or cationic.
- Polymerization of mixtures of reactants (e.g., styrene, vinylpyridine, and divinylbenzene) that can polymerize by additional polymerization. At least one of the reactants must contain an anionic or cationic moiety, or one that can be made so. Also, one of the reactants is usually a cross-linking agent to provide control of the solubility of the films in water.
- Introduction of anionic or cationic moieties into preformed films by techniques such as imbibing styrene into polyethylene films, polymerizing the imbibed monomer, and then sulfonating the styrene. A small amount of crosslinking agent (e.g., divinylbenzene) may be added to control leaching of the ion-exchange component. Other similar techniques, such as graft polymerization of imbibed monomers, have been used to attach ionized groups onto the molecular chains of preformed films.
- Casting films from a solution of a mixture of a linear film-forming polymer and a linear polyelectrolyte, and then evaporating the solvent.

Membranes made by any of the above methods may be cast or formed around scrims or other reinforcing materials to improve their strength and dimensional stability.

Table 1 shows the properties of some representative commercially available ion-exchange membranes as reported by the manufacturers.

III. CONCENTRATION AND DEPLETION BY ELECTRODIALYSIS WITH ION-EXCHANGE MEMBRANES

Figure 1 illustrates the principle of electrodialysis. An electrodialysis unit consists of a number of thin compartments through which solutions containing dissolved electrolytes are pumped. These compartments are separated by alternate cation- and anion-exchange membranes, which are spaced about 1 mm, or less, apart. The end compartments contain electrodes.

When electric current passes through the solution compartments and membranes, cations tend to migrate toward the negatively charged electrode (cathode) and anions tend to migrate toward the positively charged electrode (anode). The cations and anions in one set of solution compartments can pass freely through the cation- and anion-exchange membranes that are the walls of this first set of compartments. However, once the cations and anions are in the second set of solution compartments (the alternate compartments) cations are blocked from further transfer because the anion-exchange membranes will not allow their passage, and anions are blocked similarly from further transfer because they are blocked by cation-exchange membranes. An ion-depleted solution can be withdrawn from the first set of compartments; an ion-enriched solution can be withdrawn from the second set of compartments.

IV. CONCENTRATION POLARIZATION AND WATER-SPLITTING

The major limitation of the production rates achievable in electrodialysis units is concentration polarization at the surfaces of the ion-exchange membranes. Concentration polarization occurs because of differences in the transport numbers of ions in solutions and in the ion-exchange membranes. To illustrate the source of concentration polarization that occurs at an anion-exchange membrane, consider the fact that the transport number of anions in a solution is lower than it is in an anion-exchange membrane. Because of the lower transport number, the number of negative ions transported through the solution to the surface of an anion-exchange membrane by the electrical current is not sufficient to make up for the negative ions removed from that surface and transferred through the membrane. This deficiency of ions results in a reduction of the concentration of ions in the solution at the surface of the membrane. Eventually a concentration gradient is established in the solution such that the balance of the ions needed for steady-state operation (i.e., those not supplied by the electical current) are supplied by the diffusional transport resulting from the concentration gradient.

At the other surface of an anion-exchange membrane accumulation of ions occurs because more ions are transferred through the membrane than can be carried away by the electric current (because the transport numbers are lower in solution than in the membrane). Thus, the concentration of ions in the solution at the surface of the membrane increases, and a concentration gradient is established in the solution that results in removal of the excess ions by diffusion.

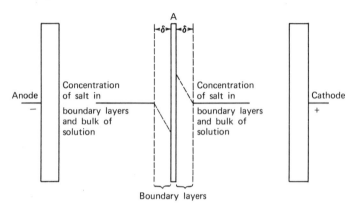

Figure 2. Concentration gradients that occur in boundary layers near membrane surfaces. A, anion-exchange membrane; δ, thickness of idealized boundary layer.

In Figure 2, concentration gradients of the types just described are depicted. In an electrodialysis compartment, the solutions flow past the membranes and a continuous velocity gradient extends from the surfaces of the membranes to the center of the flow channel. The flow channels are filled with spacer materials that cause complex flow patterns. The velocity of the solutions past the membranes and through the spacer materials results in relatively good mixing of the solution in the center portions of the flow channels, but the mixing is less near the surfaces of the membranes. Near the surfaces, boundary layers of almost static solution exist. In Figure 2, the so-called Nernst idealized model* of boundary layers has been depicted for simplicity in illustration.

* In the Nernst idealized model of boundary layers, it is assumed that complete mixing of the solution in the center of the flow channels occurs, and that there is a sharp line of demarcation between completely mixed solution and the completely static solution in boundary layers adjacent to the membrane surfaces.

Concentration gradients as shown by the dashed lines in Figure 2 are established in the static boundary layers such that the concentration of ions at the solution–membrane interface on the side of the membrane that ions enter is lower than it is in the completely mixed zone. The concentration of ions at the interface on the other side of the membranes is higher than that in the completely mixed zone.

As the electric current density is increased these interfacial concentrations become lower (on the entering side) and higher (on the other side). As the current density is increased still more, the concentration of ions at the entering interface approaches zero. At this current density (termed the *limiting* current density) H^+ and OH^- ions formed by ionization of water begin to be conducted through the solution and through the membrane. The OH^- ions transferred through the membrane can cause deleterious changes of pH in the membrane and in the solutions in the boundary layers. Also, the layer of almost deionized water in the boundary layer on the entering side causes a large increase in resistance. Moreover, at current densities in excess of the limiting current density, little additional transfer of the desired ions is obtained because they are not available at the entering side. The increment of current above the limiting values results mostly in conduction of H^+ and OH^- ions from the so-called splitting of water, and in only a small amount of additional transport of the desired ions. Thus concentration polarization imposes a real limitation on the production rate (i.e., transport of desired ions) in an electrodialysis unit.

A detailed mathematical analysis of concentration polarization that relates the thickness of the boundary layers to the limiting current density is given in Chapter 2. The effects of velocity on the thickness of the boundary layers and on the limiting current density are also discussed in Chapter 2.

V. PROBLEMS AND LIMITATIONS OF CONVENTIONAL ELECTRO-DIALYSIS

Several problems have been encountered with the conventional electrodialysis process illustrated in Figure 1, in which both cation-exchange and anion-exchange membranes are used. Many of the problems are associated with the anion-selective membranes.

When the limiting current density is exceeded, the hydroxyl ions that are produced at anion-exchange membranes transfer through anion-exchange membranes and cause the concentrating solutions to become basic. If the concentrating solutions contain materials that form precipitates in basic solutions, such as $Ca(HCO_3)_2$, the concentrating compartments can become blocked by the precipitates and the electrodialyzers then become inoperative.

Such troubles with precipitating solids have been overcome in some instances by acidification of the concentrating solutions.

A more serious problem is encountered with the anion-exchange membranes that have been available up to 1970, when solutions containing large organic anions are treated. Organic anions driven into the matrix of the anion-exchange membranes by the electrical driving force become almost irreversibly attached to the surfaces of the positively charged anion-exchange membranes if the organic portion of the anion is large. Once this attachment occurs, the surface layer (i.e., the attached anions) of what had been a positively charged anion-exchange membrane acquires a negative fixed charge and acts like a cation-exchange membrane. The result of these changes is that the resistance of the membranes increases and the selectivity decreases.

Because the operating lifetimes of anion-exchange membranes are often short when solutions with organic anions are treated, a variant of electrodialysis called transport depletion in which cation-exchange and neutral membranes are used has been developed to avoid the difficulties with fouling of anion-exchange membranes by organic materials.

Research is currently underway* to develop anion-exchange membranes with relatively porous matrices and with matrices based on aliphatic polymers instead of aromatic polymers. Either, or both, of these efforts may result in anion-exchange membranes that are less susceptible to organic fouling than the relatively tight, aromatic-based membranes in use now.

Although there are a few problems with conventional electrodialysis, such as those discussed above, there are only two serious limitations on the use of the process:

- The degree of concentration possible is limited by the water transferred through the membranes by osmosis and electroosmosis.
- The concentration of salts in the product water that is achievable with low energy costs is about 200–400 ppm of electrolytes because of the high cost of transferring ions through the high electrical resistances of solutions that are more dilute.

Despite these limitations electrodialysis, with its several variants, is a useful process for a number of industrial separations.

VI. VARIANTS OF ELECTRODIALYSIS

Several variants of the basic electrodialysis process have been developed to overcome some of the problems that have been encountered with the basic process.

* Supported by the Office of Saline Water of the U.S. Department of the Interior.

A.   Electrodialysis with Cation-Exchange and Neutral Membranes

In one of the variants of electrodialysis suggested by Deming and studied by Lacey[11] and by Lang and Huffman,[12] an array of alternate cation-exchange and nonselective neutral membranes is used, as shown in Figure 3. When

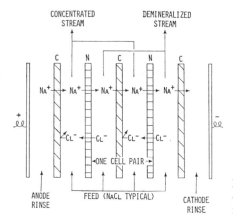

Figure 3.   The cation-neutral transport depletion process. C, cation-selective membrane; N, neutral membrane.

direct electric current is passed through an assembly of this type, depleted and concentrated boundary layers form at the two sides of the cation-exchange membranes, but there is no such effect at the nonselective neutral membranes. The neutral membranes serve only to separate the depleted and concentrated boundary layers from each other and, in doing so, create alternate diluting and concentrating compartments.

The principal advantages of this electrodialysis process over the conventional electrodialysis process stem from the elimination of anion-exchange membranes and the attendant polarization and fouling problems. This process, therefore, permits greater flexibility in the selection of flow rates and current densities for a given application. The neutral membranes may also serve to screen solutes on the basis of molecular size and thereby contribute to a multiple-separation process. The use of this cation-neutral membrane process for demineralizing whey is discussed in Chapter 4.

B.   Electrogravitational Demineralization and Concentrations

The electrogravitational process is an adaptation, in which ion-exchange membranes are used, of the much older process of electrodecantation discovered by Pauli[13] in 1924 and applied commercially to rubber-latex separations by Stamberger[14] in 1939. The term electrophoresis-convection was applied to this type of separation by Kirkwood[15] in 1941. A comprehensive treatise on electrophoresis was prepared by Bier.[16] The term electrogravitation appears to have been suggested by Murphy[17] in 1950 because of the analogy with thermogravitation, and Frilette[18] used the term electrogravitational transport when he reported on its applicability to demineralization in 1957. Kollsman[19] has described some improvements on this technique and its application to other chemical separations.

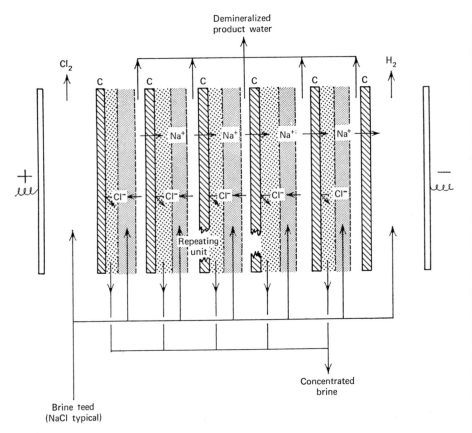

Figure 4. Electrogravitation with cation-selective membranes.

Electrogravitational demineralization may be achieved in a cell such as that depicted in Figure 4, in which only cation-exchange membranes are used. When a direct electric current is passed through a system of this type, depleted and concentrated boundary layers form at each side of the membrane just as they do in the previously described electrodialysis processes, but the solution in the depleted boundary layer rises and collects at the top of the cell because it has a lower specific gravity than that of the bulk of the solution, and the solution in the concentrated boundary layer slides downward and collects at the bottom of the cell. During the continued passage of current, the density of the depleted layer is further reduced by electrical heating, which results from its relatively high electrical resistance. Conversely, the concentrated boundary layer has low electrical resistance and is subjected to the least electrical heating. Convective separation is thus augmented by the heating effects resulting from the passage of current. In operation, a depleted solution is withdrawn from the tops of the cells, and a concentrated solution is withdrawn from the bottoms of the cells, both at flow rates that do not disturb the convective flow within the cells.

One of the advantages of this process is the extreme simplicity of construction. The membranes may be placed loosely or simply hung in a tank fitted with electrodes, as shown by Murphy.[20] The intermembrane compartments need not be isolated from each other with expensive frames or gaskets; imperfections in the membranes, even holes, will not seriously affect performance. The direction of electrical current may even be reversed without altering the locations at which depleted and concentrated solutions are removed. The pumping-energy requirements for this process are virtually nil because of the low flow rates usually employed, and maintenance of an electrogravitational cell is easy because of the simplicity.

Studies of the electrogravitational demineralization process by Mintz and Lang[21] have shown that concentration gradients of greater than 100 to 1 can be achieved in cells only 6 in. high. Studies by Lang and Huffman[12] showed that electrogravitation is not competitive with other processes for demineralizing saline water but may be of interest for some industrial separations.

C. Electrosorption

In another variant of electrodialysis studied by Lacey and Lang[22,23] under the sponsorship of the Office of Saline Water, three-layer membranes comprised of a neutral inner layer (which may be merely solution) sandwiched between a cation-exchange layer and an anion-exchange layer are used. Many of these three-layer membranes can be arranged between a pair of electrodes so that the solution to be treated flows between the membranes. No gaskets or solution manifolds are needed. The three-layer membranes are called electrosorption membranes. One method of forming them is to treat a flattened plastic tube (cellulosic dialysis tubing, for example) to impart anion-exchange properties to one face and cation-exchange properties to the other face. During the passage of a direct electric current, the solution on the outside of the electrosorption membranes is depleted and the solution within the membrane becomes highly concentrated, as illustrated in Figure 5. The concentrated brine is subsequently "unloaded" from the membranes by reversing the direction of current. If the tops of such envelopelike membranes are provided with a means of overflow, the electrosorption process is very

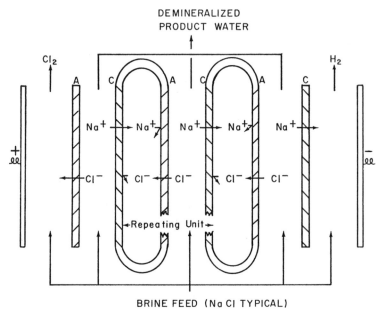

Figure 5. The electrosorption process.

much like the unit-cell type of electrodialytic salt concentrator developed by Japanese firms (see Chapter 6). The main advantages of electrosorption (or unit-cell salt concentrators) are the high percentage of the membrane area utilized, the simplicity, and the low cost of the stacks.

D. Combination Electrodialysis Processes

The variants of electrodialysis described above can be combined in different ways to utilize some of the advantages that each possess. The BALC process for recovering pulping chemicals from sulfite pulping liquors that is described in a later chapter (Chapter 5) makes use of combinations of neutral and ion-exchange membranes.

As another example, an electrodialysis process in which cation-exchange and neutral membranes are used might be operated with countercurrent flow of the depleted and concentrated streams at flow rates similar to those used in electrogravitation. Lang and Huffman[12] showed that the neutral membrane aids in the separation of the depleted and concentrated boundary layers and permits higher throughput rates than are possible with an all-cation membrane assembly. Anion/neutral and all-anion membrane counterparts of cation/ neutral and all-cation membrane systems are, of course, also possible.

In addition to the process variants that result from different membrane arrangements and internal flow conditions, most of the various external process-stream handling techniques that are used for conventional electrodialysis may be applied to its variants. Thus, batch recirculation with and without mixing, feed-and-bleed, and continuous-flow systems warrant consideration for any particular application, as shown by Mintz.[24] The various methods of handling process streams (e.g., batch recirculation and feed-and-bleed) are discussed in detail in Chapter 3.

VII. TYPES OF SEPARATIONS POSSIBLE BY ELECTRODIALYSIS

Each of the electrodialysis processes described above may be applied to a large variety of separation problems. Some typical classes of separations that are possible by electrodialysis are discussed briefly below.

A. Demineralization

Since the fundamental mechanism of electrodialysis is the transport of electrolytes through membranes, it is apparent that electrodialysis processes offer the opportunity for separation of ionized from nonionized or weakly ionized components in solution. The demineralization of sugar solutions and whey solutions are examples of this type of application.

Demineralization of nonaqueous solutions, such as alcohol and liquid ammonia, appears possible on the basis of membrane properties measured in these solutions, although the long-term stability of ion-selective membranes in such media has not been demonstrated.

B. Concentration of Electrolytes

Concentration of electrolytes by electrodialysis usually accompanies demineralization, and the degree of concentration is limited only by the extent to which solvent transport accompanies ion transport. In the production of concentrated brine from seawater, for example, it has been difficult to achieve concentrations higher than about 3.5 N because of the water transported through the membranes with the ions. Precipitation of insoluble salts may also be a limitation in some concentration processes, but studies of the electrosorption process by Lacey[22] have shown that some precipitates can be handled successfully by this process.

C. Ion-Replacement Reactions

Continuous anion replacement or continuous cation replacement may be achieved in an electrodialysis system with all anion-exchange membranes or all cation-exchange membranes, respectively. Every second compartment is fed with the solution to be treated, and the intervening compartments are fed with a solution containing a relatively high concentration of the desired ion.

One example of continuous anion replacement is the sweetening of citrus juices as described by Smith.[25] In this process, hydroxyl ions furnished by a caustic solution replace the citrate ions in the juice. The hydroxyl ions are neutralized by hydrogen ions in the juice to produce water and are, therefore, unavailable for further transfer through the membrane system. The overall result is the removal of citric acid from the juice.

An example of continuous cation replacement is the removal of radioactive strontium from milk. In this process, a synthetic electrolyte solution having a cationic composition similar to milk furnishes nonradioactive cations to the milk solution to replace radioactive strontium.

Continuously controlled pH adjustment of solutions without the direct addition of acids or bases is another possible application of continuous ion replacement by electrodialysis.

D. Metathesis Reactions

If the diluting compartments of a conventional electrodialysis system with alternating anion- and cation-exchange membranes are fed with electrolyte solutions of different compositions, double decomposition reactions are readily achieved. For example, KNO_3 and $NaCl$ can be readily produced from KCl and $NaNO_3$. Photographic emulsions of $AgBr$ have been produced by this technique from $NaBr$ and $AgNO_3$, with a gelatin emulsion being fed into the concentrating compartments. The emulsion thus produced is free of extraneous electrolytes.

E. Separation of Electrolysis Products

When the products of electrolysis at one electrode or the reactants fed to one electrode must be kept separated from the reactants or products at the other electrode, ion-selective membranes sometimes offer an advantage over porous separator materials. For example, ion-exchange membranes have been used as cell separators in a process for hydrodimerizing acrylonitrile to adiponitrile.[26]

F. Fractionation of Electrolytes

In addition to separating electrolytes from solutions on the basis of their charge sign, membrane systems can be used to separate ions of like charge on the basis of their different rates of transport through membranes. When natural brackish waters are demineralized in most electrodialysis systems, for example, calcium and magnesium ions are transported more rapidly than sodium ions. The demineralized water is thereby softened to a greater degree than would be expected from the proportional reduction of all ionic components.

Ionic fractionation can also be achieved in an electrogravitation system, the more mobile ions concentrating at the bottom of the cell and the least mobile ions concentrating at higher levels in the cell, as shown by Kollsman.[27]

REFERENCES

1. P. H. Prausnitz and J. Reitstotter, *Electrophorese, Electroosmose, Electrodialyse*, Stenkopff, Dresden, 1931.
2. E. Manegold and C. Kalauch, *Kolloid-Z.* **86**, 93 (1939).
3. K. H. Meyer and W. Strauss, *Helv. Chim. Acta.* **23**, 795 (1940).
4. J. R. Wilson, *Demineralization by Electrodialysis*, Butterworth, London, 1960.
5. S. B. Tuwiner, *Diffusion and Membrane Technology*, Reinhold, New York, 1962.
6. K. S. Spiegler, *Principles of Desalination*, Academic, New York, 1966.
7. N. Lakshminarayanaiah, *Chem. Revs.* **65**, 494 (1965).
8. H. Z. Friedlander and R. N. Rickles, *Anal. Chem.* **37**, 27A (1966).
9. F. Helferrich, *Ion Exchange*, McGraw-Hill, New York, 1962.
10. R. E. Lacey, Ed., *Membrane Processes for Industry*, Southern Research Institute, Birmingham, Ala., 1966.
11. R. E. Lacey, *U.S. Off. Saline Water Res. Dev. Rep.* **80** (1963).
12. E. W. Lang and E. L. Huffman, *U.S. Off. Saline Water Res. Dev. Rep.* **439** (1969).
13. W. Pauli, *Biochem. Z.* **152**, 355 (1924).
14. P. Stamberger and E. Schmidt, Brit. Pats. 505,752 and 505,753 (May 15, 1939).
15. J. G. Kirkwood, *J. Chem. Phys.* **9**, 878 (1941).
16. M. Bier, *Electrophoresis*, Academic, New York, 1959.
17. G. W. Murphy, *Electrochem. Soc.* **97**, 405 (1950).
18. V. J. Frilette, *Phys. Chem.* **61**, 168 (1957).
19. P. Kollsman, U.S. Pats. 2,854,393 (Sept. 30, 1958) and 3,099,615 (July 30, 1963).
20. E. A. Murphy, F. V. Paton, and J. Ansell, U.S. Pat. 2,331,494 (October 12 ,1943).

21. M. S. Mintz and E. W. Lang, *Study of Electrogravitational Separation*, Southern Research Institute Report 7374-1650-I (1965).

22. R. E. Lacey and E. W. Lang, *U.S. Off. Saline Water Res. Dev. Rep.* **106** (1964).

23. R. E. Lacey and E. W. Lang, *U.S. Off. Saline Water Res. Dev. Rep.* **398** (1969).

24. M. S. Mintz, *Ind. Eng. Chem.* **55,** 18 (1963).

25. R. N. Smith, C. T. Hicks, and R. J. Moshey, *The AMF Process for Citrus Juice*, American Machinery and Foundry Co., Research Div., Springdale, Conn.

26. M. M. Baizer, *Tetrahedron Letters* **15,** 973 (1963).

27. P. Kollsman, U.S. Pat. 3,025,227 (March 13, 1962).

Chapter II   Physiochemical Aspects of Electromembrane Processes

Thomas A. Davis* and George F. Brockman†

The objective of this chapter is to provide the reader with theoretical information that is needed for an understanding of electromembrane processes. The topics to be discussed include electrolytic conduction, the nature of ion-exchange materials, equilibria between ion-exchange materials and solutions, and the driving forces and resistances encountered in electromembrane transport.

This chapter also includes a discussion of the hydrodynamics in electromembrane devices and the relationships between solution velocities and the thicknesses of boundary layers that exist at the surfaces of membranes (which sometimes constitute one of the controlling resistances in ion transport). The last topic discussed in the chapter is the phenomena that occur at electrodes in electromembrane processes.

I. NATURE OF ION-EXCHANGE MATERIALS

The first commercially important ion-exchange materials were the zeolites. These natural mineral crystals are composed of an aluminosilicate lattice containing fixed anionic charges to which the cations are bound by ionic attraction. It was found in the nineteenth century that these cations are free to exchange with cations in a solution in contact with the zeolites, and that this exchange is reversible.

* Southern Research Institute, Birmingham, Alabama.
† University of Alabama at Birmingham.

Although natural and synthetic zeolites are still used commercially, synthetic ion-exchange resins are much more widely used. In these resins, the insoluble matrix is a crosslinked polymer to which are attached charged radicals such as $-SO_3^-$ or $-NR_3^+$. The most commonly used matrix material is a copolymer of styrene and divinylbenzene, but other matrix materials, such as polyethylene and fluorocarbons, are also used. Other fixed ionizable radicals that are used include $-COO^-$, $-PO_3^{-2}$, $-HPO_2^-$, $-AsO_3^{-2}$, $-SeO_3^-$, and various tertiary and quaternary amines. Resins are tailored to meet specific demands as to color, size, shape, durability, exchange capacity, and selectivity. Ion-exchange resins are manufactured in the forms of beads and membranes, and at least one manufacturer extrudes fluorocarbon-base resins as monofilaments and tubes.

The most valuable property of ion-exchange beads is their ability to exchange ions with solutions and to be regenerated by exchange in the reverse direction. For instance, a solution of NaCl may be passed through a bed that contains beads of two resins, one resin containing exchangeable H^+ ions, the other resin containing exchangeable OH^- ions. The beads will exchange ions for the Na^+ and Cl^- ions in the solution, and the effluent water will have a low—often very low—NaCl content. Ion-exchange resin beads are often used in cyclic processes in which the exchange in the reverse direction is performed in a separate regenerating step so that the beads can be used over and over. A large amount of information is available in the literature[1-3] and from manufacturers of ion-exchange resins[4] concerning the selection and use of ion-exchange beads, but ion-exchange membranes are the main topic of interest here.

Ion-exchange membranes are usually used in continuous processes. The most valuable property of ion-exchange membranes is their permselectivity, which is the ability to allow the passage of certain ionic species while preventing the passage of other species. Because this property of the membranes is more important than their exchange capacity as such, they are often called permselective or ion-selective membranes.

Because ion-exchange beads and membranes are both made of ion-exchange resins, they exhibit similar chemical and physical behavior. However, many aspects of ion-exchange that are important in membranes are irrelevant in beads and vice versa. We deal here primarily with the principles of ion exchange, as they relate to the operation of electromembrane processes.

A. Models for Ion-Exchange Membranes

Tuwiner[5] defines a membrane as a material or device that acts as a physical barrier between two fluids while permitting a degree of communication between them. Helfferich[1] classifies ion-exchange membranes as homogeneous or heterogeneous, depending on their visible structure. Heterogeneous membranes are membranes that are composed of more than one material. Typical heterogeneous membranes are made by grinding ion-exchange beads and mixing the ion-exchange material with a binder. This mixture is cast or calendered onto a support fabric that imparts strength and dimensional stability to the membrane. Homogeneous membranes are of uniform composition (except, of course, on the molecular scale), and the physical properties of the membranes are essentially the same from point to point.

Of the theoretical models that have been used to describe ion-exchange membranes, the most popular have been homogeneous models. Mackay and Meares[6] present the following six assumptions as the basis for a homogeneous model of cation-exchange membranes:

1. The membrane consists of an isotropic three-dimensional network of polymer chains to which anionic groups are chemically bound. This is swollen by water or an electrolyte solution which constitutes an internal aqueous phase.

2. The anionic groups are completely ionized, and are uniformly distributed through the polymer.

3. The internal aqueous phase contains free cations (termed counterions because their charges are opposite to the charges affixed to the membrane matrix) to balance the electrical charge of the fixed ions. Sorbed electrolyte, made up of an equal number of free cations and anions, may also be present. The internal aqueous phase is qualitatively similar to an ordinary aqueous electrolyte solution.

4. The fixed anions are surrounded by, or project into, the internal aqueous phase. The polymer network to which these ions are fixed has the thermal vibrations and rotations characteristic of a swollen rubberlike gel, and there is a definite volume concentration of fixed ions in the system in any particular state.

5. The average distance between junction points of the rubberlike polymer network is large compared to the dimensions of the free ions in the internal solution. Therefore, the network exerts no mechanical sievelike effect on the movement of the free ions.

6. The volume concentration of fixed ions does not vary appreciably with the electrolyte concentration in the external solution with which the membrane is in contact.

Since ion-exchange membranes contain an appreciable concentration of mobile ions, they display good electrolytic conductivity, usually about the same as that of 0.01–0.05 N salt solutions.

II. EQUILIBRIUM IN ION-EXCHANGE SYSTEMS

The most important principle involved in ion-exchange membrane processes is the Donnan equilibrium relationship. The development of this relationship is treated in detail by Helfferich.[1] For a dilute solution of uni-univalent electrolyte, such as NaCl, it can be shown that the concentration of the co-ion in the membrane phase is

$$\bar{C}_{\text{co-ion}} = \left(\frac{C^2_{\text{co-ion}}}{\bar{M}_R}\right)\left(\frac{\gamma_\pm}{\bar{\gamma}_\pm}\right)^2 \tag{1}$$

where the overbar refers to the membrane phase, $\bar{M}_R$ is the concentration of fixed charges per unit volume of solvent in the membrane, and $\gamma_\pm$ is the mean activity coefficient of the salt in the designated phase. In a typical situation in which a membrane with a fixed ion concentration of 5×10^{-3} eq/cm^3 of internal solvent ($\bar{M}_R = 5 \times 10^{-3}$) is immersed in a solution of 0.01 N NaCl ($C = 1 \times 10^{-5}$), the co-ion concentration in the membrane is only 2×10^{-8} eq/cm^3 of solvent.

In the above calculation it was assumed that the ratio of activity coefficients, $\gamma_\pm/\bar{\gamma}_\pm$, is unity. Although activity coefficients generally approach unity in dilute solutions, there is considerable evidence that this is not necessarily the case within the membrane.[1] Osmotic swelling, specific interactions of the ions with the fixed charges or other groups on the resin matrix, and effects of size and charge of the ions all combine to make resins of a given composition more selective to some ionic species than to others. For example, there are commercially available membranes that will selectively transport univalent ions and reject (or partially reject) multivalent ions in electrodialysis. With these membranes, not only the concentrations but also the compositions of electrolyte solutions can be altered by electrodialysis. (See Chapter 6.) Detailed discussion of such effects is beyond the scope of this book, but readers who wish to pursue these subjects further may refer to Helfferich[1] and Lakshminarayanaiah.[7]

Equation 1 describes the phenomenon known as Donnan exclusion. For a cation-exchange system with a dilute external solution, the anions are repelled by the fixed charge of the membrane to such an extent that the concentration of anions in the pore liquid is proportional to the square of the concentration in the external solution. The cations, on the other hand, may move freely throughout the pore liquid in the membrane. They are required only to satisfy conditions of electroneutrality. The concentration of cations in the pore liquid of a cation-exchange membrane resin is almost independent of the concentration of the external solution. With concentrated external solutions co-ion intrusion into the membrane increases the co-ion transport number of the membranes and reduces the electrical resistance because the internal solution is more concentrated.

The Donnan equilibrium that exists between the pore liquid in an ion-exchange resin and an external solution has two aspects of importance to electromembrane processes. For a cation-exchange membrane, first, compared to the external solution, the cation concentration is high and the anion concentration is low in the pore liquid, and second, cation-exchange resins in the form of membranes can be used as semipermeable membranes, which suppress the transfer of anions while allowing the transfer of cations.

III. ELECTROLYTIC CONDUCTION

When a salt, such as sodium chloride, is dissolved in water it dissociates into positively and negatively charged particles called cations and anions, respectively. Cations and anions move in opposite directions in an electric field. In electrolytic conduction, electrical charges are transported through solutions by the motion of these ions.

The velocity, u, at which a given ionic species, i, is moved by an electrochemical driving force or potential gradient in the x direction is shown by the generalized equation

$$u_i = -k \frac{dE}{dx} \tag{2}$$

where k is a proportionality constant, the magnitude of which is determined by the properties of the ion and the medium through which it is driven, and E is the potential.

The electrochemical potential is the sum of the chemical and the electrical potentials. If the potential is expressed in electrical units (volts), the velocity is proportional to the ionic mobility, m_i cm²/(V)(sec), and the ionic charge, or valence, Z_i.

$$u_i = -m_i Z_i \frac{dE}{dx} \tag{3}$$

If the potential is expressed in chemical terms, the proportionality factor is the diffusion coefficient, D cm²/sec:

$$u_i = -D_i \frac{d \ln C_i}{dx} \quad \text{(Fick's first law)} \tag{4}$$

The ionic mobility and the diffusion coefficient are related by the Nernst-Einstein equation,

$$m_i = \frac{FD_i}{RT} \tag{5}$$

where R is the gas law constant, T is absolute temperature, and F is Faraday's constant.

In most applications we are concerned with fluxes rather than velocities. The flux, J_i, is the product of the velocity, u_i cm/sec, and the concentration, C_i mole/cm³.

$$J_i = u_i C_i \text{ mole}/(\text{cm}^2)(\text{sec}) \tag{6}$$

Ionic flux is related to the electric current density, i, by Faraday's constant, $F = 96{,}500$ A-sec/eq, and the valence, Z_i eq/mole.

$$i = F \sum_i Z_i J_i \text{ A/cm}^2 \tag{7}$$

Collisions between moving ions and essentially stationary species such as solvent molecules, or other moving species such as oppositely charged ions, dissipate energy. The degree of dissipation of energy is characterized by resistance to the flow of electrical current. It is often convenient to express relationships between voltages and currents in terms of conductivity, κ, which is the reciprocal of the resistivity.

Conductivity data are often presented in terms of equivalent conductance. The equivalent conductance, Λ, is defined in terms of conductivity and concentration as

$$\Lambda = \frac{\kappa}{C} = 1000 \frac{\kappa}{N} \text{ cm}^2/(\text{ohm})(\text{eq}) \tag{8}$$

where C is the concentration of the solution in eq/cm³, and N is the normality in eq/l.

A useful term is the equivalent ionic conductance, λ_i, which is proportional to the ionic mobility.

$$\lambda_i = m_i F \text{ cm}^2/(\text{ohm})(\text{eq}) \tag{9}$$

The equivalent conductance, Λ, of an electrolyte is the sum of the ionic conductances of its ions.

$$\Lambda = \lambda_+ + \lambda_- \text{ cm}^2/(\text{ohm})(\text{eq}) \tag{10}$$

The values of equivalent ionic conductance reported in the literature usually pertain to solutions at infinite dilution. Equivalent conductance decreases with increasing concentration because of nonidealities in the solutions. For highly dissociated electrolytes such as NaCl the decrease is moderate, but for highly associated electrolytes such as acetic acid, a drastic reduction in equivalent conductance occurs. Equivalent conductance increases with temperature at a rate of about $2\%/C°$. The precise value of the temperature factor depends upon the nature of the electrolyte and its temperature and concentration. According to Craig,[8] actual values range from 0.1 to about $9\%/C°$.

The relative movement of the different ions in a solution is denoted by transport numbers. The transport number of a given ion in a solution is simply the ratio of the electric current conveyed by that ion to the total current.

$$t_i = \frac{Z_i J_i}{\sum_i Z_i J_i} = \frac{m_i Z_i^2 C_i}{\sum_i m_i Z_i^2 C_i} \tag{11}$$

Transport numbers are positive for all ions. With knowledge of the concentrations and mobilities (or equivalent ionic conductances) of the ions in a solution, the transport number of any species in the solution can be determined. For a solution of a uni-univalent salt such as NaCl, the transport numbers are

$$t_+ = \frac{m_+}{m_+ + m_-} \quad \text{and} \quad t_- = \frac{m_-}{m_+ + m_-} \tag{12}$$

IV. FACTORS AFFECTING ENERGY CONSUMPTION

One question that should be asked when electrodialysis is being considered as a means of removing electrolytes from a solution is how much electrical energy will be consumed in removing a given amount of electrolyte from a given solution. The factors affecting energy consumption are summarized in this section and treated in more detail in the following sections.

The amount of salt removed from a depleting stream of an electrodialysis cell pair* is proportional to the current density; and the proportionality factors are the coulomb efficiency, η, and Faraday's constant, F.

$$\text{electrolyte removal rate} = \eta i/F \text{ eq}/(\text{cm}^2)(\text{sec}) \tag{13}$$

The coulomb efficiency is related to the transport numbers of the membranes,

$$\eta = 1 - (t_-^c + t_+^a) \tag{14}$$

in which t_-^c is the anion transport number of the cation-exchange membrane and t_+^a is the cation transport number of the anion-exchange membrane.

---

* An electrodialysis cell pair is the repeating unit of an electrodialysis stack. It consists of an anion-exchange membrane, a cation-exchange membrane, an enriching-solution compartment, and a depleting-solution compartment.

For perfectly semipermeable membranes $t_-^c = t_+^a = 0$, but in practice co-ion transport numbers in the range 0.01–0.05 are common for commercially available ion-exchange membranes in dilute solutions.

The actual coulomb efficiency in an electrodialysis stack is lower than that calculated from the transport numbers of the membranes due to loss of current through solution manifolds and diffusion of electrolyte through the membranes. Nevertheless, coulomb efficiencies of 0.9 or greater are often achieved in electrodialysis of dilute solutions.

The voltage drop per cell pair is proportional to the current density and the cell-pair resistance, R_{cp} ohm-cm^2. Typical electrodialysis stacks operate with voltage drops of about 2 V/cell pair.

The energy consumption can be expressed as

$$\text{energy consumption} = \frac{iR_{cp}F}{\eta} \text{ W-sec/eq removed} \tag{15}$$

Thus the energy consumption increases with current density and cell-pair resistance and decreases with improved coulomb efficiency. Typical energy consumption of about 50 W-hr/eq removed can be expected in electrodialysis of dilute solutions.

V. CONCENTRATION POLARIZATION IN ELECTROMEMBRANE PROCESSES

In almost all electromembrane processes the solutions to be treated flow between parallel planar ion-exchange membranes. (See Figure 1 of Chapter 1.) Thus the hydrodynamics in such devices is that of flow between parallel plates.

The hydrodynamics of fluids flowing past flat plates has been discussed by Lamb,[9] Rouse,[10] and Levich[11] among others, and the effect of solution flow on the performance of electrodialyzers has been reported by Mandersloot,[12] Cooke,[13] Rosenberg and Tirrell,[14] Cowan and Brown,[15] Weiner, Rapier, and Baker,[16] Gunter,[17] and Davis and Lacey[18] among others. The discussions presented here are based on an idealized model (i.e., the so-called Nernst[19] idealization) in which the following are assumed:

- There are boundary layers adjacent to the membranes in which the solutions are completely static.
- The solution in the interior of a solution compartment (i.e., between the boundary layers) is thoroughly mixed so that the concentration of electrolyte at any point in this zone is the same as that at any other point.
- There is no change of either the thickness of the boundary layers or the concentration gradients along the flow channel.

The actual concentration and velocity profiles in electrodialysis compartments approach parabolic functions more nearly than the discontinuous profiles of the Nernst model. Moreover, there are entrance effects (regions near the entrances to solution compartments in which the velocity and concentration profiles are developing).[20] Nevertheless, the Nernst model affords a simplified approach to mathematical developments, which results in expressions that are easy to use and that predict performance adequately for use in the design of electromembrane processes.

Concentration Gradients in Boundary Layers. In electromembrane processes, the transport numbers of ions in the solutions are not equal to those in the membranes. With the passage of electric current through such a system, ions tend to become concentrated in certain locations and depleted in others. This tendency for concentration and depletion is opposed by diffusion and physical mixing.

The flux of cations through a Nernst boundary layer as a result of electrical transport is

$$[J_+]_e = t_+ \frac{i}{FZ_+} \tag{16}$$

The flux of cations through a cation-exchange membrane is

$$[\bar{J}_+]_e = \bar{t}_+ \frac{i}{FZ_+} \tag{17}$$

Since $t_+ < \bar{t}_+$ in a cation-exchange membrane, the flux of cations through the solution up to or away from a cation-exchange membrane as a result of electrical transport is less than the flux of cations through the membrane. On the side of the membrane that cations enter, there is depletion of ions, and concentration effects result in the establishment of a concentration gradient in the boundary layer such that the ions needed for steady-state transport through the cation-exchange membrane are supplied by diffusion.

The flux of ions resulting from diffusion can be expressed in terms of Fick's first law:

$$[J_\pm]_d = -D \frac{dC}{dx} \tag{18}$$

At steady state, the combined electrical and diffusive flux through the solution in the boundary layers equals the electrical flux through the membranes, and the electrical flux through the membrane is essentially the total flux (neglecting diffusion in the membrane, which is usually negligibly small).

$$\bar{J}_+ = \bar{t}_+ \frac{i}{FZ_+} = -D \frac{dC}{dx} + t_+ \frac{i}{FZ_+} \tag{19}$$

Integration of Equation 19 yields a simple relationship between boundary layer thickness, δ, current density, i, interfacial concentration, C_0, and the so-called bulk concentration, C_b.

$$C_0 = C_b \pm (t_+ - t_+)i\,\frac{\delta}{DFZ_+} \tag{20}$$

Whether the concentration, C_0, at a given interface is greater or less than the bulk concentration is determined by the relative values of the transport numbers and the direction of electrical current flow.

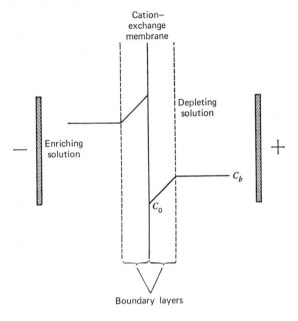

Figure 1. Concentration gradients in electrodialysis.

Concentration gradients in boundary layers are illustrated in Figure 1. The cation-exchange membrane in the center is highly permeable to cations (i.e., t_+ approaches unity) and almost impermeable to anions (i.e., t_- approaches zero). Cations are transferred through the membrane faster than they can arrive from the solution on the right or leave to the solution on the left by electrical transfer. Thus the concentration at the right interface decreases and the concentration at the left interface increases until concentration gradients are established such that diffusion provides the additional fluxes needed to maintain steady-state conditions in the boundary layers.

If the current is raised high enough, the concentration at the right interface will approach zero. Since there will not then be enough salt ions to carry the current, any further current through the cation-exchange membrane will be carried by hydrogen ions formed by the ionization of water at the interface or by enforced co-ion intrusion from the opposite side, or by both mechanisms. The current at which the interfacial concentration becomes essentially zero at some point in the apparatus is called the limiting current density, because any further increase in current density would cause loss of coulomb efficiency, greatly increased electrical resistance, and changes in the pH of the solutions, all of which are detrimental to the operation of electromembrane devices.

The limiting current density is described by setting $C_0 = 0$ in Equation 20, in which case

$$i_{lim} = \frac{C_b DFZ_+}{\delta(t_+ - t_+)} \tag{21}$$

The value of the Nernst boundary layer thickness, δ, is determined by the hydraulics of the system, but since it is a characteristic of an idealized system, its value cannot be predicted accurately from hydraulic data alone. Values of δ are usually obtained by determining experimentally the value of i_{lim} and calculating δ from Equation 21. Once the value of δ has been determined for given hydraulic conditions in a particular apparatus, it is useful for estimating limiting current densities for feed solutions of varying concentrations. The value of δ can be decreased (with a resulting increase in i_{lim}) by increasing the temperature or the velocity of the solutions through the compartments of the electrodialysis stack. An increase in temperature also results in an increase in the diffusivity, D, thus an increase in the limiting current density.

It is evident that, all else being constant, if the bulk concentration, C_b, is increased, a higher current density can be used before the concentration will decrease to zero at the membrane surface. In fact, the limiting current density is approximately proportional to the bulk concentration. The bulk concentration varies along the direction of flow in an electrodialysis compartment, and in most systems the current also varies. Cowen and Brown[15] showed that for a simple cell with thick compartments the limiting current density theoretically varies with the log-mean concentration* in the depleting compartment. The quantity, i/N_{lm}, (sometimes written i/N or CD/N) is called the polarization parameter. Once the limiting polarization parameter, i_{lim}/N_{lm} has been determined for a system, it can be used to estimate the limiting current density for the same apparatus operating with solutions of other concentrations because it is nearly independent of concentration.

_____

* The log-mean concentration for this application is defined as $(C_f - C_p)/(\ln C_f/C_p)$, where f and p refer to the feed and the depleted product.

VI. ELECTRICAL ENERGY REQUIREMENT

The major energy requirements for ionic transport in electromembrane processes arise from the need to overcome the electrical resistances of the solutions and the membranes, and the back electromotive forces (membrane potentials and diffusion potentials) caused by concentration gradients.

The electrical resistance of a solution-filled compartment Δ cm thick is

$$\rho_{soln} = \Delta/C\Lambda \text{ ohm-cm}^2 \tag{22}$$

where C is the solute concentration in eq/cm^3 and Λ is the equivalent conductance in $cm^2/(ohm)(eq)$.

The resistance of a membrane cannot be calculated by a simple method but it can be measured. Although direct current is used in electromembrane processes, alternating currents are usually used to measure the electrical resistance of membranes because concentration gradients that are present with direct current systems are not formed with alternating current and the resistance of the membrane itself can be more easily determined. However, the resistance of a membrane to alternating current is lower than the resistance to direct current. Spiegler[21] has discussed some theoretical reasons for the difference in resistance. A preliminary estimate of the resistance of a membrane to direct current can be made by measuring its resistance to alternating current in a simple bridge circuit[22] and multiplying the result by

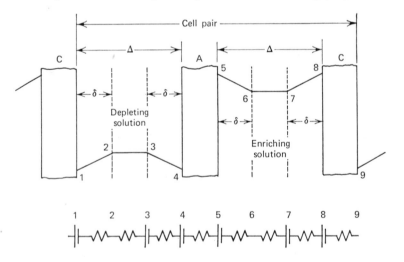

Figure 2. Simplified concentration profiles in an electrodialysis cell pair and the analogous electrical circuit.

1.75, but if precise values are needed the resistance to direct current should be determined under the conditions of membrane use.

As current flows through an electrodialysis apparatus, concentration gradients and discontinuities are established. Figure 2 shows schematically the concentration profiles in an electrodialysis cell pair. The cell pair contains the concentrated and depleted bulk solutions, the two membranes, and the four boundary layers in which the concentrations of salts in the solution near the membranes vary considerably.

The electrical circuit in Figure 2 is an analog of the electrodialysis cell pair. The battery symbols represent the concentration potentials, and the resistor symbols represent the resistances of membranes and segments of solution. Lacey[23] employed this analog to develop an equation for the total potential across a cell pair. The voltage from 1 to 2 is the algebraic sum of the diffusion potential and the IR drop from point 1 to point 2*

$$E_{1-2} = \frac{RT}{F} \int_1^2 (t_d^- - t_d^+) \, d \ln \gamma C + i \int_1^2 \frac{dx}{C\Lambda} \tag{23}$$

where the subscript d refers to the depleting solution.

The other individual voltages (e.g., from 2 to 3 and from 3 to 4) can be similarly expressed. The algebraic sum of the individual voltages from 1 to 9 is the applied voltage for the electrodialysis cell pair. (Note that the conditions at point 9 are identical to those at point 1.)

$$
\begin{aligned}
E_{1-9} = \frac{RT}{F} & \left[(t_d^- - t_d^+) \ln \frac{\gamma_4 C_4}{\gamma_1 C_1} + (t_a^- - t_a^+) \ln \frac{\gamma_5 C_5}{\gamma_4 C_4} \right. \\
& + (t_e^- - t_e^+) \ln \frac{\gamma_8 C_8}{\gamma_5 C_5} + (t_c^- - t_c^+) \ln \frac{\gamma_9 C_9}{\gamma_8 C_8} \right] \\
& + i \left[\delta \frac{\ln C_1 \Lambda_1 / C_2 \Lambda_2}{(C_1 \Lambda_1 - C_2 \Lambda_2)} + \frac{\Delta - 2\delta}{C_2 \Lambda_2} + \delta \frac{\ln C_3 \Lambda_3 / C_4 \Lambda_4}{(C_3 \Lambda_3 - C_4 \Lambda_4)} \right. \\
& \left. + \rho_a + \delta \frac{\ln C_5 \Lambda_5 / C_6 \Lambda_6}{(C_5 \Lambda_5 - C_6 \Lambda_6)} + \frac{\Delta - 2\delta}{C_6 \Lambda_6} + \delta \frac{\ln C_7 \Lambda_7 / C_8 \Lambda_8}{(C_7 \Lambda_7 - C_8 \Lambda_8)} + \rho_c \right]
\end{aligned}
\tag{24}
$$

The subscripts d and e refer to the depleting and enriching solutions and a and c refer to the anion- and cation-exchange membranes. The numerical subscripts refer to the corresponding numbers in Figure 2.

---

* The expression derived here is for a system containing a uni-univalent electrolyte. It is assumed that the transport numbers of the membranes and solutions do not change over the range of concentrations of interest. Also, the assumption is made that single-ion activities can be replaced by molalities and mean activity coefficients. If other than uni-valent ions are present, the single-ion activities must be properly converted to molalities and activity coefficients as discussed by I. M. Klotz.[24]

Although Equation 24 is cumbersome and requires the assumption of some terms for which exact values will not be known, it has been shown that reasonably accurate predictions of the total potential drop across a cell pair can be made with this equation. Use of Equation 24 is time-consuming with manual calculation methods, but it is easily programmed for computers. It can be simplified by making several assumptions, most of which can be approached in practice. The simplifying assumptions are as follows:

- The transport numbers of anions and cations in the solution are equal, $t_s^- = t_s^+$. (This is essentially true for potassium chloride.)
- The equivalent conductance is independent of concentration over the limited range of interest, $\Lambda_1 = \Lambda_2 = \cdots = \Lambda$.
- The membranes are ideally permselective, $t_c^+ = t_a^- = 1, t_a^+ = t_c^- = 0$.
- The activities are equal to the concentrations, $\gamma_1 = \gamma_2 = \cdots = 1$.

With the assumptions above the concentration profiles become symmetrical, $C_1 = C_4, C_5 = C_8$, and Equation 15 simplifies to

$$E_{1-9} = 2\frac{RT}{F}\ln\frac{C_5}{C_4} + i\left[2\frac{\delta}{\Lambda}\left(\frac{\ln C_1/C_2}{(C_1 - C_2)} + \frac{\ln C_5/C_6}{(C_5 - C_6)}\right)\right.$$
$$\left. + \frac{\Delta - 2\delta}{\Lambda}\left(\frac{1}{C_2} + \frac{1}{C_6}\right) + \rho_a + \rho_c\right] \quad (25)$$

When the current through the stack is alternating or is an infinitesimal direct current, Equation 25 is reduced to

$$E_{1-9} = i\left[\frac{2\Delta}{\Lambda C} + \rho_a + \rho_c\right] \quad (26)$$

because all concentrations in the cell pair are equal and no concentration gradients exist.

VII. ELECTRODE REACTIONS

Since electromembrane processes require the passage of an electric current that must enter the apparatus from an external circuit, a brief discussion of the reactions that occur at the electrodes is in order. Generally, the electromembrane processes are operated with direct current and the anode and cathode are distinct, although in certain processes, the current may be reversed periodically. The technology of electrode reactions is highly developed in many respects, but there is still considerable controversy concerning mechanisms and the relative importance of competing reactions that occur at electrodes. Our discussion is limited to pointing out some of the problems.

The cathode or negatively charged electrode is the source of electrons, and at the cathode, the electrons must be transferred from the external circuit to ions in the solution. The following are typical reactions by which this transfer of charge may be accomplished:

$$M^{+x} + xe^- \rightarrow M^0, \text{ metal deposition}$$

$$O_2 + 2H_2O + 4e^- \rightarrow 4OH^-, \text{ reduction of gaseous oxygen}$$

$$2H^+ + 2e^- \rightarrow H_2 \text{ (acidic solution)} \qquad \text{evolution}$$
$$\qquad\qquad\qquad\qquad\qquad\qquad\qquad\qquad \text{of gaseous}$$
$$2H_2O + 2e^- \rightarrow H_2 + 2OH^- \text{ (basic solution)} \qquad \text{hydrogen}$$

Metal-deposition reactions are useful in processes such as electroplating and the recovery of spent pickle liquor. Reactions in which gaseous oxygen is reduced are important in fuel cells. Reactions in which hydrogen is evolved at the cathode with no change in the composition of the electrode are commonly encountered in electromembrane processes such as electro-dialysis. There is usually almost no deterioration of the cathode, and almost any conductor that is compatible with the rest of the system can be used as a cathode. Carbon steel is a commonly used cathode material.

Selection of a suitable anode material is a problem because the anode is subjected to severe oxidation. Depending on the composition of the solution, pH, anode composition, and current density, one of more of the following reactions may occur at the anode:

$$M^0 \rightarrow M^{+x} + xe^-, \text{ metal dissolution}$$

$$H_2 \rightarrow 2H^+ + 2e^-, \text{ oxidation of gaseous hydrogen}$$

$$2H_2O \rightarrow O_2 + 4H^+ + 4e^- \text{ (acidic solutions)} \quad \text{evolution}$$
$$4OH^- \rightarrow O_2 + 2H_2O + 4e^- \text{ (basic solutions)} \quad \begin{array}{l}\text{of gaseous}\\ \text{oxygen}\end{array}$$

$$2Cl^- \rightarrow Cl_2 + 2e^-, \text{ evolution of gaseous chlorine}$$

$$M^0 + xOH^- \rightarrow M(OH)_x + xe^- \qquad \begin{array}{c}\text{oxidation}\\ \text{of}\end{array}$$
$$2M^0 + 2xOH^- \rightarrow M_2O_x + xH_2O + 2xe^- \qquad \text{electrode}$$

Metal dissolution results in destruction of the electrode, and may be minimized by the use of noble metals, such as platinum, for the anode. Oxidation of gaseous hydrogen is an important reaction in fuel-cell operation. Reactions in which gaseous oxygen or chlorine is evolved are commonly encountered in electromembrane processes and in the electrolytic manu-facture of these gases.

When the anode is oxidized, hydroxyl ions are consumed. Unless pro-vision is made for removing the companion hydrogen ions or supplying hydroxyl ions, the electrode solution will become acidic. Most metal oxides and hydroxides are soluble in acidic solutions.

$$M_2O_x + 2xH^+ \rightarrow 2M^{+x} + xH_2O$$

or

$$M(OH)_x + xH^+ \rightarrow M^{+x} + xH_2O$$

The net result is the dissolution of electrode metal. Durable electrodes can be made from noble metals such as platimun, but the costs are generally prohibitive. Thin coatings of platinum on certain metals, such as titanium or tantalum, have proven satisfactory in some cases.[25] Oxides of some metals such as lead and ruthenium have proved to be sufficiently conductive and insoluble in acids to be used as coatings for anodes.[26,27] Magnetite electrodes have been used for anodes in electrodialysis, but this material is very fragile. Although it deteriorates with use, graphite is fairly inexpensive, easy to machine, and for some uses its oxidation products do not contaminate the electrode solutions. Therefore, graphite is often used as a material for anodes.

REFERENCES

1. F. G. Helfferich, *Ion Exchange*, McGraw-Hill, New York, 1962.
2. R. Kunin, *Ion Exchange Resins*, Wiley, New York, 1958.
3. S. B. Applebaum, *Demineralization by Ion Exchange*, Academic, New York, 1968.
4. Diamond Shamrock Chemical Company, *Duolite Ion-Exchange Manual*, 1969.
5. S. B. Tuwiner, *Diffusion and Membrane Technology*, Reinhold, New York, 1962.
6. D. Mackay and P. Meares, *Proc. Roy. Soc. (London) A* **232**, 498 (1955).
7. N. Lakshminarayanaiah, *Transport Phenomena in Membranes*, Academic, New York, 1969.
8. D. N. Craig, *J. Res. Natl. Bur. Stand.* **21**, 225 (1938).
9. H. Lamb, *Hydrodynamics*, Dover, New York, 1945.
10. H. Rouse, *Elementary Mechanics of Fluids*, Wiley, New York, 1946.
11. V. C. Levich, *Physiochemical Hydrodynamics*, Prentice-Hall, Englewood Cliffs, N.J., 1962.
12. W. G. B. Mandersloot and R. E. Hicks, *Ind. Eng. Chem. Process Des. Dev.* **4**, 304 (1965).
13. B. A. Cooke, *Proceedings, First International Symposium on Water Desalination*, Vol. 2, U.S. Dept. of Interior, 1965, p. 219.
14. N. W. Rosenberg and C. E. Tirrell, *Ind. Eng. Chem.* **40**, 780 (1957).
15. D. A. Cowan and J. H. Brown, *Ind. Eng. Chem.* **51**, 1445 (1959).
16. S. A. Weiner, P. H. Rapier, and W. K. Baker, *Ind. Eng. Chem. Process Des. Dev.* **3**, 126 (1964).
17. G. A. Gunter, *U.S. Off. Saline Water Res. Dev. Rep.* **459** (1966).

18. T. A. Davis and R. E. Lacey, "Forced-Flow Electrodesalination," *U.S. Off. Saline Water Res. Dev. Rep.* **557** (1970).

19. W. Nernst, *Z. Physik Chem. (Leipzig)* **47**, 52 (1904).

20. A. A. Sonin and R. F. Probstein, *U.S. Off. Saline Water Res. Dev. Rep.* **375**, (1968).

21. K. S. Spiegler, *Interim Rep. No. 4, U.S. Off. Saline Water Grant No. 14-01-0001-652*, 13 (1966).

22. "Test Manual for Permselective Membranes," *U.S. Off. Saline Water Res. Dev. Prog. Rep.* **77**, 156 (1964).

23. R. E. Lacey, "Desalination by Electrosorption and Desorption," *U.S. Off. Saline Water Res. Dev. Rep.* **228**, 82 (1967).

24. I. M. Klotz, *Chemical Thermodynamics*, Prentice-Hall, New York, 1950, p. 309.

25. Anonymous, *Chem. Eng.* **77** (19), 36 (1970).

26. R. Thangappan et al., *Ind. Eng. Chem. Prod. Res. Dev.* **9** (4), 563 (1970).

27. Anonymous, *Chem. Eng.* **72** (15), 82 (1965).

Chapter III   Engineering and Economic Considerations in Electromembrane Processing

Everett L. Huffman* and Robert E. Lacey*

In this chapter, some of the more important engineering and economic aspects of electromembrane processes are discussed. No claim is made for complete coverage of these topics, since such a task would require more space than is available. However, an attempt has been made to provide the reader with basic information about the engineering and economic aspects of electromembrane processes and to provide references in which the reader may find more detailed discussions of the topics.

I. BASIC REQUIREMENTS OF ELECTROMEMBRANE PLANTS

A. Flow Diagram of Electrodialysis Process

A flow diagram for a two-stage electrodialysis plant for concentrating or depleting electrolytes in solution is shown in Figure 1. Solution to be fed to the concentrating compartments of the electrodialysis stacks (assumed to be free of gross particulate matter) is stored in the concentrate-feed tank. It is pumped from the feed tank through a filter and into the concentrating compartments of the first-stage electrodialysis stack, and from there through the concentrating compartments of the second-stage stack. The flow path for the solution to be depleted of electrolytes is similar, except that the depleting

* Southern Research Institute, Birmingham, Alabama.

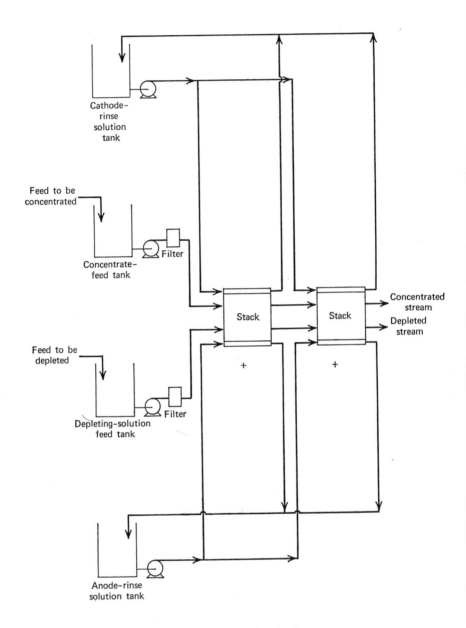

Figure 1. Flow diagram for a two-stage electrodialysis plant.

stream is pumped through the depleting compartments of the first- and second-stage electrodialysis stacks. In the electrodialysis stacks, ions are transferred through the cation- and anion-exchange membranes, as described in Chapters 1 and 2, to effect depletion of the electrolytes in the solution in one set of compartments and concentration of the electrolytes in the alternate compartments.

Anode- and cathode-rinse solutions, which may be the solutions being treated or specially prepared solutions of various kinds, are pumped from the electrode-rinse solution tanks through the electrode compartments of the stacks, and returned to the proper tank. The pH of each of these solutions is usually maintained at a value that will counteract the bases produced by electrochemical action at the cathodes or the acids produced at the anodes, so that H^+ or OH^- ions do not enter the concentrating and depleting compartments of the electrodialysis stacks. The concentrated and depleted streams leaving the second-stage stack may be pumped to product storage tanks, if the desired amount of concentration or depletion is achieved, or to further stages of electrodialysis, if further concentration or depletion is desired.

In addition to the conventional electrodialysis process, there are several other electromembrane processes that make use of various combinations of nonselective membranes, bipolar electrodes, and cation- and anion-exchange membranes. The flow schemes used with some of these processes are described in other chapters of this book, such as that for cation-neutral transport depletion in Chapter 4 and for the BALC process in Chapter 5.

B. Materials of Construction

The materials used in electromembrane stacks, pumps, piping, tanks, and other equipment associated with electromembrane processes are usually chosen on the basis of the requirements for corrosion resistance, physical strength, and resistance to electrochemical attack by the solutions to be treated. Corrosion data for many materials that may be used in electromembrane processing are given in the *Biennial Materials of Construction Reports* published by *Chemical Engineering* magazine. For applications in which conventional electrodialysis stacks are used, satisfactory materials for construction are usually chosen by the manufacturers of electrodialysis equipment. For applications in which special electromembrane stacks are used (such as those with bipolar electrodes, or special arrangements of membranes), the stack components may have to be developed to meet specific requirements. In such instances, data on the materials used in conventional electrochemical equipment, such as chlor–alkali cells, hydrogen–oxygen cells, and hydrogen peroxide cells, have been found to be helpful. Such data can be found in certain textbooks and handbooks[1,2] as well as in the chemical literature. Information from manufacturers of reinforced plastic materials and electrode materials has also been found to be helpful.

Piping that is not electrically conducting, such as poly(vinyl chloride) and reinforced epoxy-resin pipe, is often used to transfer solutions to and from the electromembrane stacks. Such piping also can offer advantages in corrosion resistance.

Plastics, both reinforced and unreinforced, are often the material of choice for tanks, although plastic-lined low-carbon steel tanks have also been used.

C. Pretreatment of Feed Solutions

Provision should be made to pretreat the solutions fed to electromembrane stacks to remove particulate matter and to prevent bacterial growth within the stacks when treating solutions that will support bacterial growth. Particulate matter tends to deposit within the solution compartments. Any deposits increase the resistance of cell pairs and interfere with the proper distribution of solutions within the compartments. In extreme instances, such deposits can block or seriously reduce solution flow to compartments. Reduction of solution flow can result in inadequate mixing within the affected solution compartments with attendant difficulties with polarization (see Chapter 2 for discussion of the effects of solution velocity on polarization).

For applications in which the solutions being treated can support bacterial growth, either continuous chlorination of the feed solutions or other bactericidal treatment is necessary, or the stack must be periodically treated to destroy bacterial growth within it.

For removal of particulate matter the filtration requirements depend on the size and shape of the particles. Colloidal and gelatinous particles and "stringy" particles give particular trouble because they adhere more tenaciously to the meshlike spacers within the solution compartments than do granular particles. It is usually good practice to remove particulate matter as completely as possible.*

For applications in which the feed solutions contain small amounts of tri- or tetravalent cations, such as iron, manganese, or thorium, in addition to the main ions that are to be transferred through the membranes, the multivalent ions should be removed. It has been found that multivalent cations can attach themselves to the fixed negative charges within cation-exchange membranes in such a way as to partially, or completely, neutralize the

---

* In work within the authors' organization, electromembrane systems have been operated successfully with feed solutions containing colloidal organic material that resulted in turbidities as high as 30 Jackson turbidity units, by periodically cleaning the stacks with solutions containing enzyme-active agents, or with solutions of very high ionic strength (e.g., concentrated NaCl) that would deagglomerate colloids.

negative charges. This neutralization of charge results in reduced values of transport numbers for other cations with an attendant decrease in coulomb efficiencies. DeKorosy[3] found that not only charge neutralization but even charge reversal could occur with tetravalent ions. Apparently, the tetravalent positive ions attached themselves to the negative fixed charges in a manner that left at least one of the four positive charges un-neutralized by fixed negative charges so that the membrane matrix eventually acquired a net positive charge instead of a negative charge. The formerly cation-exchange membranes then became anion-exchange membranes.

It has also been found that when small amounts of iron and manganese are present in the feed solutions and the pH is sufficiently high, insoluble hydroxides of iron and manganese form on and in anion-exchange membranes, and increase their resistances unless such ions are sequestered.*

Because of these difficulties, traces of ions such as iron and manganese should be removed from the feed solutions before they are introduced into electromembrane stacks.

Iron and manganese may be removed from most feed solutions by oxidation, adjustment of pH, and filtration, or by pretreatment with manganese zeolite and filtration. The concentration of these ions in feed solutions should not be greater than 0.05 ppm. It may be desirable to remove traces of some heavy metals by treatment with H_2S or Na_2S and filtration.

D. Pumps and Typical Hydraulic Pressure Drops

The primary considerations in choosing pumps for use in electromembrane processing are corrosion resistance, introduction of undesirable metallic contaminants, and the hydraulic pressure drops through the equipment. The resistance of materials to corrosion in use depends on the nature of the solutions to be treated in the particular application. The corrosion-resistance requirements for pumps are no different from those for other chemical processes. If pumps are chosen to be satisfactory on the basis of corrosion resistance, they will usually not introduce undesirable amounts of heavy metals into the feed solutions.

The hydraulic pressure drops encountered in electrodialysis equipment will depend on the design of the equipment used. Typical pressure drops through tortuous-path units are from 30 to 60 psi. Pressure drops through sheet-flow units are much lower—from 3 to 10 psi. (See Section II of this chapter for descriptions of tortuous-path and sheet-flow electrodialysis units.)

* In an electromembrane process for regenerating sulfuric acid from spent pickling acids, Calman and Heit (U.S. Pat. 3,394,068) found that sequestration of iron ions with triethanolamine eliminated the deposition of iron hydroxides within anion-exchange membranes.

E. Storage Tanks

Minimum sizes of interstage tanks (if used) depend on the time needed to shut the entire system down when a tank level is too low, or on the time needed to start a pump when a tank level is too high. Feed tanks and interstage tanks should be chosen to be large enough to ensure a supply of feed solutions equal to at least a few minutes of pumping capacity for the pumps involved. The sizes of feed tanks may also depend on surges, shutdowns, and other features of the process furnishing the feed. Sizes of product-storage tanks depend on the inventory of stored product desired, which is dependent on factors other than the operation of the electromembrane process itself.

F. Power Supply

D-c energy for electromembrane stacks is usually supplied with transformers and rectification equipment, although motor-generator sets have been used on occasion. Power supplies suitable for use may be made with single-phase or three-phase transformers and silicon diodes arranged as a full-wave rectifier. Capacitors are generally used with single-phase circuits to give low-ripple d-c voltage, but with three-phase transformers and full-wave rectifiers the output voltage has acceptably small ripple without the use of capacitors.

G. Controls

The following variables are usually measured or controlled, or both:

- D-c voltage and current supplied to each electrodialysis unit.
- Flow rates and pressures of the depleting and concentrating streams, and of the electrode rinse streams.
- Electrolyte concentrations of the depleting and concentrating streams at the inlets and outlets to the electrodialysis stacks.
- pH of the depleting stream and the electrode rinse streams.

All the variables above are interrelated. Automatic control of the flows of the depleting and concentrating streams can be achieved by the use of flow-type conductivity cells in the effluent depleting and concentrating streams along with a controller that compares the conductivities of the depleting and concentrating streams with that of a preset resistance and actuates flow-control valves in the liquid supply lines.

To prevent damage to the membranes or other components of an electromembrane stack in the event of stoppage of liquid flows to the stacks, the equipment should be provided with fail-safe devices that will turn off the power to the stacks and pumps. This can be accomplished by placing flow-measuring controllers in the feed streams which will turn off the power to the stacks and pumps, if the flow rate of any stream drops below a preset value.

H. Operational Limitations of Electromembrane Processes

Excessive concentration polarization at the surfaces of the membranes can limit the current densities that can be used in electromembrane processing. The relationships between solution velocities and excessive concentration polarization have been discussed in Chapter 2.

The degree of concentration that can be achieved is limited by the amount of water that is transferred through the membranes along with the ions by osmosis and electroosmosis. The flux of water that occurs with a flux of ions is highly dependent on the nature of the membranes. The concomitant fluxes of water and ions have not been studied extensively, but Lakshminarayanaiah[4] and Lacey[5] have reported some data on the subject. In general, the number of milliliters of water transported per faraday decreases with increases in current density, decreases with increases in solution concentration, and decreases as the water content of the membrane decreases.

Another limitation on the degree of concentration that can be accomplished in some applications of electromembrane processing is that some compounds present in the feed solutions may exceed their maximum solubility if the feed is concentrated too much. For example, in the electrodialytic concentration of seawater to furnish brines for the chlor–alkali industry, the formation of precipitates of calcium sulfate in the concentrating compartments limits the degree of concentration that can be achieved. The developers of this process (see Chapter 6) solved the problem by developing ion-exchange membranes that allowed the transfer of Na^+ and Cl^- ions in preference to Ca^{2+} and SO_4^{2-} ions.

In addition to the technical limitations above there is an economic limitation on the degree of depletion that is feasible in some applications of electromembrane processes. As the concentration of electrolyte in the depleting stream decreases, the electrical resistance of the solution increases, I^2R losses increase, and, at some point, an excessive amount of energy becomes necessary to effect additional depletion of the solution and the high costs for energy cause the process to become noncompetitive with other types of processing.

II. ELECTRODIALYSIS STACKS

Most of the equipment used in electromembrane processes are standard items commonly used in the process industries, such as pumps, tanks, or filters, but the membrane stack is unique. An electromembrane stack is

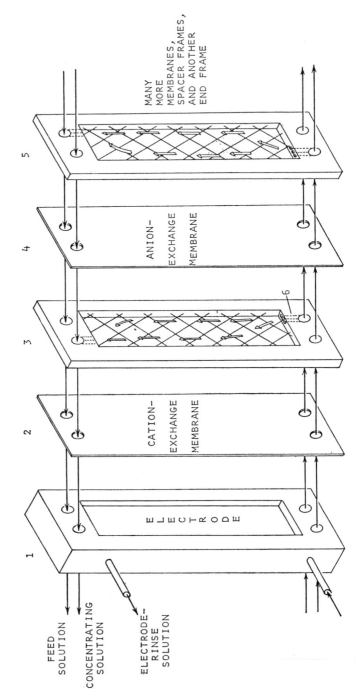

Figure 2. Exploded view of components in an electromembrane stack.

essentially a device to hold an array of membranes between electrodes in such a way that the streams being processed are kept separated.

Figure 2 is an exploded view of part of an electromembrane stack that shows the main components. Component 1 in Figure 2 is one of the two end frames, each of which has provisions for holding an electrode and introducing and withdrawing the depleting, the concentrating, and the electrode-rinse solutions. The end frames are usually made relatively thick and rigid so that pressure can be applied easily to hold the stack components together. The inside surfaces of the electrodes are recessed, as shown, so that an electrode-rinse compartment is formed when an ion-exchange membrane, component 2, is clamped in place. Components 3 and 5 are spacer frames. Spacer frames have gaskets at the edges and ends so that solution compartments are formed when ion-exchange membranes and spacer frames are clamped together.

Usually the supply ducts for the various solutions are formed by matching holes in the spacer frames, membranes, gaskets, and end frames. Each spacer frame is provided with solution channels (6 in Figure 2) that connect the solution-supply ducts with the solution compartments. The spacer frames have mesh spacers, or some other device, in the compartment space to support the ion-exchange membranes so they cannot collapse when there is a differential pressure between two compartments.

An electromembrane stack usually has many repeated sections each consisting of components 2, 3, 4, and 5, with a second end frame at the end.

A. Criteria for Design of Stacks

The final design of a stack is almost always a compromise between a number of conflicting criteria and considerations. Some criteria for the design or selection of electromembrane stacks have been given by Wilson,[6] Hicks and Mandersloot,[7] Lacey and Lang,[8] and Schaffer and Mintz.[9] Some of the general criteria to be considered in the selection or design of an electromembrane stack are discussed under three general classifications: mechanical hydrodynamic, and electrical.

1. Mechanical Criteria

1. The maximum amount of the area of the membranes should be utilized for demineralization, because the cost of membranes constitutes a large item in the total cost of an electromembrane stack. Maximum utilization of membrane area requires that the minimum possible membrane area be obscured in the edge seals and solution-distribution devices.

2. The stacks should permit quick and easy assembly and disassembly. The dimensions of the membranes should be limited to a size that can be

handled easily by one or two men. There should be as few separate components as possible and the components should be designed so that the chance for incorrect assembly is minimized.

3. The gasket material should have negligible cold flow to prevent distortion of the gaskets and the resulting misalignment of solution entrance and exit channels. The gaskets should, however, be capable of providing positive sealing with low compression force so that only a simple, low-cost means of clamping is required. The combined requirements for gaskets permit several alternatives: (a) hard gaskets, with no cold flow, if the variations in thickness are small; (b) hard gaskets with no cold flow that have resilient faces; or (c) hard gaskets with no cold flow with provision for a resilient "line-seal" type of seal similar to O-rings.

4. The solution-distribution design for each stream should ensure equal solution flow through each compartment. Some electromembrane stacks have performed well in this respect in initial operations, but later failed because of compression and cold flow of the solution-distribution devices. The solution-distribution channels should not be easily blocked by small particles in the solution being treated.

5. The spacer screens should provide closely spaced support points for the membranes so that cold flow of the membranes into the solution compartments is minimized. The spacer screens should have a high degree of open area in the direction perpendicular to the membrane faces so that utilization of membrane area is high and electrical resistance is low. The hydraulic resistance of the spacer screen should be low in the direction of solution flow to minimize the energy needed to pump the solutions through the membrane compartments.

6. The gaskets and spacer screens should be in one plane so that the membranes are not bent or distorted. This permits the use of stiff or brittle membranes, as well as more flexible membranes.

7. The end plates of the stack, which hold the membranes and spacers frames between them and transmit the force to seal the compartments against leakage of solution, should be strong and rigid. The rigidity of the end plates is especially important for wide stacks, where beam deflection or bowing of the end plates may cause insufficient clamping pressure to be applied to some parts of the assembly of membranes, spacers, and gaskets. The required rigidity may be obtained by designing the end plates to have low beam deflection, or by the use of auxiliary clamping devices on the end plates.

Hydrodynamic Criteria

1. It is necessary to have essentially equal distribution of the solutions being processed over the width of each compartment. In narrow

compartments, the solution may be introduced at one point, but in wider compartments, introduction of solutions at a single point results in stagnation in some regions. Except in narrow compartments, the solutions should be introduced at multiple points across the widths of the compartments.

2. The solution velocity should be essentially equal at all points within a compartment. To meet this requirement, no channels or spaces between spacer screens and edge gaskets can be allowed, the compartment thickness must be essentially uniform throughout a compartment, and the hydraulic resistance of the spacer screen must be essentially uniform.

3. The solution velocities should be essentially equal in all solution compartments. Good alignment of the holes that form the supply ducts at the entrance and exit ends of the compartments must be assured, and the hydraulic pressure drops in the supply ducts must be low. Also, all the solution channels between the supply ducts and the individual compartments should have identical hydraulic resistances, and the hydraulic resistances of the spacer screens in the compartments should be identical, or the resistances of the solution channels should be high relative to the resistances in the compartments so that the flows through the compartments are controlled by the resistances of the solution channels.

4. The spacer screens used to support the membranes should cause mixing of the solution from each membrane face with the bulk solution when solutions are pumped through the membrane compartments. Effective mixing of the solutions is desirable to maintain thin interfacial boundary layers at the surfaces of the membranes.

5. Low total pressure drop across the stack is desirable to minimize the energy required for pumping and to minimize internal pressure within the stacks so that bowing of the side gaskets and bulging of the compartments is minimized.

Electrical Criteria

1. The electrical leakage through the solutions in the supply ducts should be small. To achieve low electrical leakage the resistance of the solution in the supply ducts should be high and the cross-sectional area of the supply ducts should be small. This requirement is in conflict with the need for large cross-sectional areas to provide the low hydraulic resistances as discussed above under hydrodynamic considerations.

2. The solution channels between the supply ducts and the individual compartments also should have high electrical resistances. This requirement means that the cross-sectional areas of the solution channels should be small, and coincides with the desirability of relatively high hydraulic resistances

for the solution channels as discussed above under hydrodynamic considerations.

Considerations of the criteria discussed above have led to a variety of designs for electromembrane stacks.

2. Basic Types of Stacks

Most stacks may be considered to be one of two basic types: tortuous path or sheet flow. These designations refer to the type of solution flow in the compartments of the stack.

In the tortuous-path type of electromembrane stack the solution flow path is a long narrow channel as illustrated in Figure 3, which makes several 180° bends between the entrance and exit ports of a compartment. The bottom half of the spacer gasket in Figure 3 shows the individual narrow solution channels and the cross-straps used to promote turbulence, whereas the

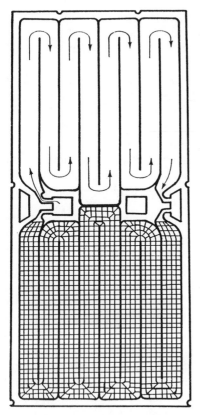

Figure 3. Diagram of a tortuous-path spacer for an electrodialysis stack.

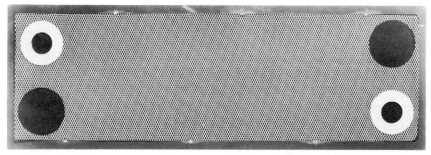

Figure 4. Photograph of a sheet-flow spacer for an electrodialysis stack.

individual channels have been omitted in the top half of the figure so the flow path could be better depicted. The ratio of the solution channel length to its width is high, usually greater than 100:1. Solution flow in a sheet-flow type of stack is approximately in a straight path from one or more entrance ports to an equal number of exit ports of a compartment, as illustrated in Figure 4. Thus the solution flows through the compartment as a sheet of liquid. The ratio of the solution channel length to width in the sheet-flow type of stack is much lower (about 1:2) than in the tortuous-flow type.

Solution velocities in the sheet-flow stacks are typically in the range of 5–15 cm/sec, whereas solution velocities are generally 30–50 cm/sec in the tortuous-path stacks. The drop in hydraulic pressure through a tortuous-path stack is generally higher than that through a sheet-flow stack because of the higher solution velocity, the changes in the direction of the flowing solution, and the longer path length.

Spacer screens are usually not needed in the solution compartments of tortuous-path stacks to support the membranes because of the narrow width of the channels and the rigid nature of the membranes used. Small straps across the solution channels provide some support for the membranes and also cause turbulence in the flowing solution. This turbulence causes mixing of the depleted or enriched solution near the membrane surfaces with the bulk of the solution and reduces the thickness of the boundary layers at the membranes surfaces.

Spacer screens are generally used in the solution compartments of sheet-flow stacks, both to support the membranes and to produce turbulence in the flowing solution.

Although some membranes can be used in either sheet-flow or tortuous-path stacks, the optimum physical or mechanical properties of membranes differ for the two types of stacks. The membranes used in tortuous-flow stacks must be sufficiently rigid to support themselves over the width of the solution flow channel. Membranes that are very flexible, or that stretch or undergo cold flow are usually not suitable for use in tortuous-path stacks, because such membranes can collapse or cold-flow into the solution-flow channels and thus prevent the attainment of equal flow rates through each compartment.

Membranes for sheet-flow stacks may be more flexible, and consequently thinner, than those for tortuous-flow stacks, because of the support provided by the spacer screens. The more rigid membranes that are normally used in tortuous-path stacks can be used in some sheet-flow stacks, but their suitability depends on the design of the solution entrance and exit ports. The entrance and exit ports of some sheet-flow stacks are not in the same plane as the rest of the membrane, and require the use of flexible membranes in order to obtain satisfactory sealing at the solution manifolds.

Sheet-flow electromembrane stacks are manufactured by several companies, including American Machine and Foundry Company, Stamford, Conn.; Asahi Chemical Industries, Ltd., Tokyo, Japan; Asahi Glass Company, Tokyo, Japan; Aqua-Chem, Inc., Waukesha, Wis.; William Boby and Company, Rickmansworth, England; Ritter-Pfaudler Corporation, Rochester, N.Y.; and Tokuyama Soda Company, Tokyo, Japan. The principal manufacturer of the tortuous-path electromembrane stacks is Ionics, Inc., Cambridge, Mass.

III. ARRANGEMENTS OF ELECTROMEMBRANE STACKS FOR VARIOUS PURPOSES

Many different arrangements of electromembrane stacks have been used to adapt the process for particular applications. Several of these arrangements and their advantages and disadvantages have been discussed by Shaffer and Mintz[9] and by Mintz.[10]

For applications in which the desired throughput or degree of demineralization is greater than is practical with one stack, combinations of stacks in parallel or in series, or both, can be used to meet the requirements. Throughput is increased in proportion to the number of stacks in parallel hydraulically. The degree of demineralization is increased progressively by adding stacks in series.

In a batch system, a fixed quantity of feed solution is circulated from a holding tank through an electromembrane stack and back to the holding tank until the desired degree of demineralization is obtained.

A feed-and-bleed system is sometimes used when large variations in the concentration of the feed solution are encountered and a continuous flow of product is desired. This system is also useful when the desired degree of demineralization is low. In the feed-and-bleed system a portion of the product solution is recirculated and blended with the raw feed solution. This blended solution then becomes the actual feed to the electromembrane stack. The production rate is the part of the product stream that "bleeds" out of the system and is not recirculated to the stack.

IV. ECONOMIC FACTORS IN ELECTROMEMBRANE PROCESSES

The individual items contributing to the total operating cost of an electromembrane process may be placed in three categories: costs that vary directly with current density, costs that vary inversely with current density, and costs that are invariant with current density. The cost of electrical energy for electromembrane processes varies directly with current density, and the costs for membrane replacement and for amortization of capital investment vary inversely with current density because less membrane area and lower capital costs are required at high current densities. Other costs, such as those for operating and maintenance labor, do not vary with current density. Figure 5 shows typical variations of these cost items with current density.

Figure 5 shows that because of the decrease of certain costs and the increase

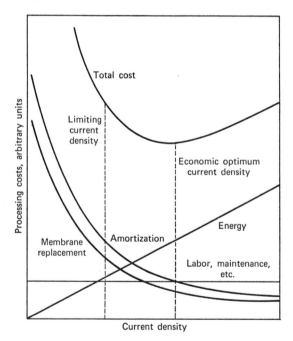

Figure 5. Variation of individual cost items making up the total processing cost.

of others as current density increases, there is an economic optimum current density for electromembrane processing. (Curves of similar nature could be developed with return on investment or payout time as the criterion of the economics of processing, so that the current densities that would result in maximum return on investment or minimum payout time could be identified.)

Cowan,[11] Tribus,[12] Lacey, Lang, and Huffman,[13] and Mattson, Snedden, and Gugeler[14] have developed equations and methods of determining the optimized costs for demineralizing saline waters. Similar methods can be used for industrial applications of electromembrane processes.

The cost-optimization method developed by Cowan involves expressing the total cost of processing as the sum of the costs of energy, membrane replacement, and amortization, and of costs that do not vary with current density. Each of these items of cost is expressed as the product of a coefficient that relates cost to the area of membranes and an appropriate term that expresses the membrane area as a function of current density. By equating the differential of the expression for total cost to zero, the current density at which the total processing cost is a minimum can be calculated.

Tribus modified the approach of Cowan by introducing terms that permit the inclusion in the cost expression of the ratio of the raw-water input rate to the fresh-water output.

Lacey et al. showed the effects on total water-production costs of a number of factors that affect costs, such as the usable lifetimes of membranes, the unit cost of membranes, the resistances of cell pairs, and the cost of energy.

The type of cost analysis used by Lacey is equivalent to determining the partial derivative of total processing cost with respect to each of the variables studied.

Mattson et al. extended this partial derivative method to include more variables that affect costs, and developed a computer program to calculate and print out the total processing costs as a function of current density. They developed equations to relate individual cost factors to such factors as production rate, reductions in salt content, membrane area, unit cost of membranes, and unit cost of electromembrane stacks. The coefficients for unit costs of such items as membranes, stacks, and rectifiers were based on extensive reviews of costs of such items (as were the exponents that adjusted such costs for changes in size). Their general equations and methods are of continuing interest, but the cost coefficients they used should be updated to reflect changes in unit costs since the time of their study.

Shaffer and Mintz[9] reviewed the above analyses of costs and pointed out that the optimum current density found by such methods is an average current density, and if more than one stage of electrodialysis is needed to effect the desired degree of desalination, the current densities in each stage differ and the average current density found from the cost analysis may not correspond to the current density in any of the stages. They concluded, however, that such procedures for cost analysis as those of Mattson and Lacey are of real value in determining the relative importance of the various factors contributing to total processing costs, so that those factors that offer opportunities for effecting worthwhile reductions in costs can be identified.

REFERENCES

1. C. L. Mantell, *Electrochemical Engineering*, McGraw-Hill, New York, 1960.
2. J. H. Perry, *Chemical Engineers Handbook*, McGraw-Hill, New York, 1963.
3. F. DeKorosy, *U.S. Off. Saline Water Res. Dev. Rep.* **380** (1968).
4. N. Lakshminarayanaiah, *Chem. Revs.* **65**, 526 (1965).
5. R. E. Lacey, *U.S. Off. Saline Water Res. Dev. Rep.* **343** (1967).
6. J. R. Wilson, Ed., *Demineralization by Electrodialysis*, Butterworth, London, 1960, Chapter 6, pp. 215–274.
7. R. E. Hicks and W. G. B. Mandersloot, "The CSIR Mark III Electrodialysis Unit," *SCIR Research Report* **247**, South African Council for Scientific and Industrial Research, Pretoria, South Africa (1966).
8. R. E. Lacey and E. W. Lang, *U.S. Off. Saline Water Res. Dev. Rep.* **398** (1969).
9. L. H. Shaffer and M. S. Mintz in *Principles of Desalination*, K. S. Spiegler, Ed., Academic, New York, 1966, Chapter 6.
10. M. S. Mintz, *Ind. Eng. Chem.* **55** (6), 18 (1963).
11. D. A. Cowan, "Interaction of Technical and Economic Demands in the Design of Large Scale Electrodialysis Demineralizers," *Advances in Chem. Ser.* **27**, American Chemical Society, Washington, D.C., 1960, p. 224.
12. M. Tribus and R. Evans, "Thermo-Economic Considerations in the Preparation of Fresh-Water from Sea-Water," *Dechema Monograph.* **47**, Deutsche Gesellschaft für Chemisches Apparatewesen, e.v., Frankfurt am Main, Germany, 1962, p. 43.
13. R. E. Lacey, E. W. Lang, and E. L. Huffman, "Economics of Demineralization by Electrodialysis," *Advances in Chem. Ser.* **38**, American Chemical Society, Washington, D.C., 1963, p. 168.
14. M. E. Mattson, L. L. Snedden, and J. E. Gugeler, "Determining the Costs of an Electrodialysis Water Desalination Plant by Parametric Equations," *Proceedings of the First International Symposium on Water Desalination, Washington, D.S., October 3–9, 1965*, Vol. 3, U.S. Govt. Printing Office, Washington, D.C., p. 265.

Chapter IV   Electromembrane Processing of Cheese Whey

Richard M. Ahlgren*

The production of cheese is one of the main phases of dairy processing in the United States, and it results in the formation of large volumes of by-product whey. At the present time, it is estimated that approximately two-thirds of the 22 billion pounds of fluid whey produced in the United States annually is either dumped into sewers or wasted in low-value disposal methods, such as direct feeding to animals, or spreading on farm fields.

The manufacture of cheese is still almost identical to that of centuries ago. The mechanism is one of coagulation of certain fractions of whole milk by either enzymatic or chemical means. Natural cheeses are produced through the addition of rennet enzymes to whole cow, goat, or other types of milk in cheese-making vats.

Under controlled conditions of time and temperature, the action of the rennet enzyme on the whole milk results in the formation of curds of much of the protein and all the fat originally present in the milk. Cottage cheese is produced by precipitation of the fat and protein present in skim milk through the action of enzymes, acids, or heat.

At the present time in the United States, almost two billion pounds of cheese are produced annually, most of which is made from cow's milk. Some types are made from goat's milk. In other parts of the world milk from camels, buffalo, and reindeer is used for cheese production. The weight of whey solids produced is about equal to the weight of cheese.

Whey is a solution that contains between 5.5 and 6.5% of dissolved solids in water. The primary constituents in whey solids are as shown below.

* Aqua-Chem, Inc., Waukesha, Wisconsin.

Substances	Quantity (%)
Protein	12
Fat	1
Lactose	70–75
Ash	8–10
Lactic acid	0.1–1.0

The normal pH of whey from the manufacturer of cheddar or natural cheese is slightly lower than 7. Fresh whey taken directly from the cheese vat has a pH of about 6.5. On the other hand, whey from cottage cheese is distinctly acidic with a pH value as low as 4.5. As whey ages a bacterial conversion of lactose to lactic acid occurs. Without precautionary measures such as pasteurization, refrigeration, or addition of spoilage retardants, it is possible for the acidity of a sweet whey to rise to about 1.0% of acid in 10–12 hr.

The formation of lactic acid in souring whey depresses the pH and denatures some of the lactalbumin protein. The protein in normal milk or fresh whey is a high-molecular-weight albumin (approximately 40,000). With the downward pH shift that occurs in the souring of whey the higher-molecular-weight proteins are divided into lower-molecular-weight fractions, such as amino acids. These amino acids contain the same amount of nitrogen as the original protein. (Kjeldahl nitrogen is used as the basis for evaluation of protein quantity.) However, these lower-molecular-weight protein fractions do not have the same functional characteristics important to their use as the original lactalbumins. They do not have the texture, whippability, mouth feel, and pleasing taste of whole milk protein.

Whey provides an excellent source of protein, milk sugar, and vitamins and minerals, but in its normal form it is not considered as a completely usable food material because of the high ash or salt content of the whey. When the ionized salts in whey are substantially removed the composition of whey approaches that of human milk. With the protein in the natural, completely functional form, whey has a very broad base of potential uses in processed cheese, synthetic milk and dairy products, ice cream, toppings and whiteners, baked goods, baby foods, candies, and other uses.

The first research in the deashing or demineralization of whey was performed by the TNO group* in Holland shortly after the World War II. At that time, food sources in Europe were in short supply, and it was thought that the wider use of whey for human consumption would help alleviate the food shortage. It had been demonstrated previously that whey could be

---

* TNO is the national research organization of Holland.

partially demineralized with the use of fixed-bed ion-exchange media, but many problems existed in this operation. There were not many types of such ion exchangers available, other than some of the naturally occurring green-sands and siliceous materials. In addition the deionization of whey could result in severe pH shifts even under carefully controlled conditions, and these changes in pH would denature the protein and result in severe fouling of the ion-exchange resins. At the time of the work by the TNO group, suitable ion exchangers in membrane form were not available. It then became a challenge to develop usable membranes, as well as to develop the process.

After about three years of effort along these lines, the workers in Holland did succeed in the demineralization of whey with the classical electrodialysis process (i.e., with anion- and cation-exchange membranes) but complications were encountered with fouling of anion-exchange membranes. Because of these difficulties the work with whey demineralization was greatly reduced, and efforts were diverted toward the use of electrodialysis as a desalination and water-supply tool.

Electromembrane processing (i.e., electrodialysis, transport depletion, and other variations) provides one of the most workable schemes for removing ionized impurities from whey. In solutions containing high percentages of ionized solids, techniques such as electrodialysis and transport depletion remove these solids on a truly continuous electrochemically efficient basis. The use of fixed-bed demineralization techniques on solutions containing approximately 0.5% or more of ionized solids is not practical since extremely short resin operating cycles and much product dilution occur. Ion-exchange resins exhibit a discrete capacity for absorption of minerals before they must be removed from service and regenerated. In contrast, ion-exchange membranes have the property of continuously passing ionized materials as long as they are under the influence of an electric current and as long as ions are available at the membrane–liquid interface. In general, removing a given amount of minerals from a solution such as whey by an electromembrane technique costs about one-eighth to one-tenth as much as removing those same minerals from the same solution with a fixed-bed ion exchange.

It has been demonstrated that the electrodialysis approach is a workable, highly efficient method for removal of a portion of the ionized whey salts. In addition to the classical cation–anion permselective membrane mode of electrodialysis, the modification using cation-neutral membranes known as transport depletion has also been applied to the process. There are advantages and disadvantages to each of these processes, but both are workable, practical methods for accomplishing the end result. For example, at the present time about 20,000,000 lb/yr of demineralized whey solids are being produced for sale by electromembrane techniques (both conventional electrodialysis and transport depletion), with additional treatment plants being planned.

The classical mode of cation–anion membrane electrodialysis is shown in Figure 1. In this technique, whey at any concentration from its natural

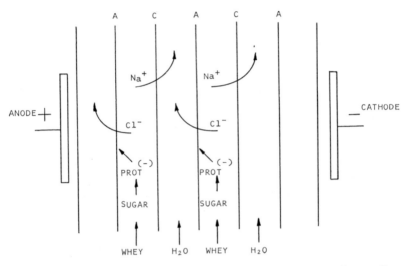

Figure 1. Electrodialysis deashing system. A, anion-exchange membrane; C, cation-exchange membrane.

concentration of about 6% solids up to more than 40–50% solids can be passed through the electrodialysis stack. The removal of ionized constituents from the whey is accomplished by transport through both anion- and cation-exchange membranes as described elsewhere in this book. Monovalent ions, primarily potassium and sodium chlorides, seem to be much more mobile than the divalent salts and are easily removed. The multivalent salts, primarily forms of calcium phosphates, seem to be much less mobile than the monovalent salts and perhaps are chelated or complexed with the protein portions of the whey. Their removal with electrodialysis can be accomplished, but generally this occurs only after the bulk of the monovalent salts has been removed, and with the application of higher electromotive driving forces.

Classical cation–anion membrane deashing of whey poses two potential problems with respect to fouling of membranes. These problems are precipitation of the less-soluble calcium minerals, and deposition of protein fractions (primarily amino acids) on anion-exchange membrane faces.

The problem of deposition of inorganic calcium phosphate scales on the surface of and within membranes can normally be controlled through measures that are used in desalination devices to control scaling, such as operation below limiting current densities, operation above minimum critical velocities, and the maintenance of proper distribution of solutions within the electrodialysis stack itself. Observation and analysis of inorganically fouled membranes from whey-processing operations indicate that a layer of calcium phosphate on the cathode side of the cation membrane is a common source of fouling. In this situation, typical gradients of ion concentrations through the electrodialysis compartments are observed. The inorganic calcium-containing deposits found on the face of the cation membrane in concentrating compartments can usually be removed by washing the membranes with acid.

In the cation–anion electrodialysis method of whey demineralization anionic protein fractions are also deposited on the faces of the anion-exchange membranes in whey compartments. Many of the denatured components of protein are large negative ions and move as such under the influence of electric current within the stack. These molecules are too large to pass through anion-exchange membranes and consequently deposit in a thin proteinaceous layer on the faces of the anion-exchange membranes in the whey compartments. Techniques such as polarity reversal can be used to dislodge these deposited materials from the membranes, but the practicality of this technique is questionable. If membranes are used continuously at operating current densities higher than approximately 20–25 mA/cm^2, this proteinaceous layer can become irreversibly deposited, resulting in rapid permanent degradation and loss of character of the anion-exchange membranes. Under typical conditions of operation with normal fresh whey at the usual operating current densities (10–25 mA/cm^2) deposition of these proteinaceous fractions usually results in membrane fouling to the point of destruction in periods of two weeks a few months. The application of cleanup and polarity-reversal techniques can sometimes extend the life of these anionic membranes for periods up to two to three times this interval. Frequent washing with high-pH cleaning solutions is an effective method of removing proteinaceous fouling.

A second mode of electromembrane processing for the demineralization of whey is with a process called transport depletion, as illustrated in Figure 2. Transport depletion is analogous to classical electrodialysis except that neutral or nonselective membranes are used instead of anion-exchange membranes. The cation-exchange membranes are identical to those used in normal electrodialysis. The nonselective membranes frequently used are dialysis membranes, or other regenerated cellulose films.

Stacks and equipment used with the transport depletion process are

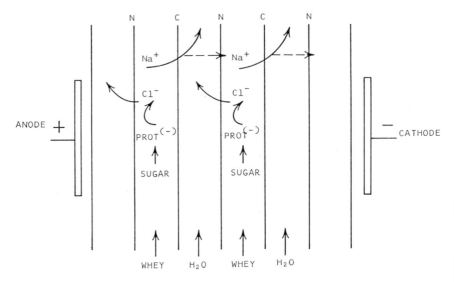

Figure 2. Transport depletion deashing system. N, neutral membrane; C, cation-exchange membrane.

identical to those required for electrodialysis. In fact, it is possible to convert a stack from electrodialysis to transport-depletion operation merely by using nonselective membranes instead of anion-exchange membranes.

The electrochemical nature of transport depletion and electrodialysis operations vary in several important aspects. The first of these differences is that concentration polarization does not occur at the surfaces of the neutral membranes used in the transport-depletion process, as it does at the surfaces of the anion-exchange membranes used in conventional electrodialysis. The result of this is that limiting current density is not approached in transport depletion. The limiting current density in any stack is almost always a function of the anion-exchange rather than the cation-exchange membranes. Polarization (and the attendant water splitting and pH shifts that result from it) causes severe denaturation of the protein in whey and increasing fouling at anion-exchange membrane surfaces. The fact that this phenomenon does not occur in the transport-depletion process makes membrane life expectancy longer and stack cleanup procedures easier with transport depletion than with normal electrodialysis. On the other hand, there is a disadvantage with transport depletion in that the attainable current efficiencies are not as high as those obtained with electrodialysis because one of the membranes is not ion-selective in nature. This lower current efficiency occurs because cations are not repelled by the neutral membranes used in transport depletion. Because cations are not repelled, they can be removed from the whey, pass through the brine or concentrate stream, and reenter the whey in the next compartment toward the cathode side. The net result is that there is a competing transport of cations in one direction and anions in the other direction through the nonselective or neutral film. The practical result of this phenomenon is that in transport depletion the attainable current efficiency is approximately one-half the potential current efficiency achieved with conventional electrodialysis that makes use of both anion- and cation-exchange membranes.

An advantage that partially offsets the disadvantage of low coulomb efficiency is that fouling of anion-exchange membranes due to proteinaceous deposits does not occur in transport depletion and therefore losses in current efficiency due to fouling do not occur. Despite the fact that conventional electrodialysis can operate at current efficiencies approaching 100% initially, there is normally a decrease in current efficiency to a level well under the 50% efficiency obtained with transport depletion because of the deposition of proteins on and in anion-exchange membranes.

In the deashing of whey (as in most food-processing applications) equipment sanitation and cleanup procedures are of vital importance. Sanitary design standards dictate that equipment should not have crevices or pockets in which particulate matter from the food stream could be caught and allowed to spoil. The basic design of electromembrane processing stacks is not compatible with this standard. However, two factors play a large role in the practical functioning of the processes. These factors are the application of cleaning-in-place (CIP) techniques and the fact that with the passage of an electric current, the amounts of bacteria in whey are normally reduced. Electromembrane stacks can be made to function on a practical sanitary basis despite the seeming incompatibility of stack design with typical sanitary construction practices.

From the electrochemical standpoint, the demineralization of whey has several specific characteristics that do not occur in other demineralization processes. Some of these features are preferential transport of monovalent ions over multivalent ions, the shifts of pH that cause protein denaturation, dialytic transport of solids, and optimum cell conductivities. The physicochemical makeup of whey imposes these variations by virtue of the buffering and chelating effects of the organic material present in the solution. Because of the relatively high viscosity of whey and the presence of organic materials that can complex simple ions, the ionic mobilities and diffusivities of electrolytes differ from the mobilities and diffusivities in simpler solutions. These differences are generally accentuated as whey is treated at higher concentrations than its normal concentration of about 6% of solids.

The inorganic salts in whey of normal composition are about one-tenth of the total dissolved solids. These salts are generally composed of about two-thirds monovalent materials, such as potassium and sodium chlorides, and about one-third divalent materials, primarily phosphates of calcium. A third ionized component of some wheys is lactic acid. The variations in lactic acid content are due to the chemistry of the original whey formation as well as the degree of souring or lactose–lactic acid conversion that has taken place. In most deashing of whey, the monovalent salts (e.g., potassium chloride) are preferentially removed as compared to the divalent salts. At the start of whey deashing operations, this monovalent preference is almost complete, but as deashing continues traces of calcium salts are also removed. In a typical whey that has been treated to remove about 65–70% of the ash, almost all the potassium chloride will have been removed, whereas less than about 20% of the calcium salts will have been taken out. Lactic acid is removed in the electromembrane processing of whey at a rate comparable to the rate of monovalent salt removal.

Like many other solutions of food substances, whey is extremely pH sensitive. The normal pH of sweet whey is about 6.5. Acid or cottage-cheese wheys may have values of pH as low as 4.5. Shifts of pH away from the isoelectric point tend to denature protein. On the acidic side of the iso-electric point, this denatured protein is agglomerated as an insoluble material. On the basic side of the isoelectric point, protein is also denatured, but under these alkaline conditions protein fractions are quite soluble. In fact, it is this solubilizing of protein under conditions of high pH that is frequently used to clean fouled membranes. Shifts of pH that accompany polarization and water splitting at membrane interfaces, when limiting currents are exceeded, frequently result in deposition of thin layers of denatured protein on membrane faces.

Electromembrane treatment of highly concentrated solutions, such as whey, poses problems with dialysis transport of dissolved solids. A solution with more than 6% dissolved solids separated from nearly pure water with a semipermeable membrane inevitably results in some dialytic transport of dissolved solids. In whey demineralization, this mechanism can result in the loss of traces of sugar from the whey to the waste-salt concentrating stream. A complicated situation exists when whey is demineralized and dialytic losses must be balanced with optimum solution conductivities. If the concentration of ionized salts in the waste brine stream is kept low to minimize ionized solids movement due to dialysis, higher electrical power consumption results because of lower conductivity. In general practice, a balance of solution concentrations in which the whey solution and the waste-salt solution have approximately the same electrical conductivity seems to give best results.

Electrical conductivity of whey is a function of three parameters: ionized salt concentration, total organic concentration, and solution temperature. Because of the high concentration of organics in whey the addition of dissolved salts results in only small increases in solution conductivity. The control of electromembrane systems in which only conductivity measurements are used is satisfactory only if conditions of temperature and overall solution concentration are maintained fairly constant. However, with the benefit of operating experience, conductivity can be used as one of the tools for control of stack operations.

The mechanics of equipment arrangement and control of a typical whey-processing operation are shown in the flow diagram in Figure 3. Because of the slightly viscous nature of high concentrations of whey solution and because the net liquid volume throughput in whey-processing plants is actually quite small, internal recirculation within stacks is normally used. The maintenance of a high velocity within stack cells is beneficial, to avoid problems with critical velocities and limiting current densities, as well as to keep particulate matter flushed from membrane surfaces. A relatively large whey-processing plant turning out about 2,000,000 lb of treated solids/yr at a concentration of about 30% by weight would be handling a liquid volume of only 2000 gal/day. Because of this fact, a feed-and-bleed system with internal recycle is usually the most practical. This method permits operation of the plant under fairly constant chemical and electrical conditions and is probably the easiest method of controlling the process.

Batch processing could also be used to treat whey. However, in the batch-processing method operating parameters are always in a transitory state and longer whey-holding time during processing is required. This longer holding time of whey at the temperature and operating conditions of the stack frequently results in an increase in bacterial counts in the product. On the other hand, the relatively low retention time of feed-and-bleed operations in which whey is taken directly from refrigerated storage, electrodialyzed, and returned to refrigerated storage usually results in lower bacterial counts in the dialyzed product than in the whey feed. It is theorized that the mechanism of this decrease in bacterial counts is due to the killing of some of the bacteria as a result of the passage of electrical current.

The processing of a relatively high-resistance fluid such as whey normally results in heat generation and a temperature rise in the product. An electro-membrane stack is a reasonably good heat exchanger, and with the use of another heat exchanger in the auxiliary stream it is possible to keep the whey product at acceptable temperature levels.

The costs of operation of electromembrane processing of whey include six individual costs. Five of these six are direct out-of-pocket expenses proportional to the amount of product produced, whereas the sixth, capital

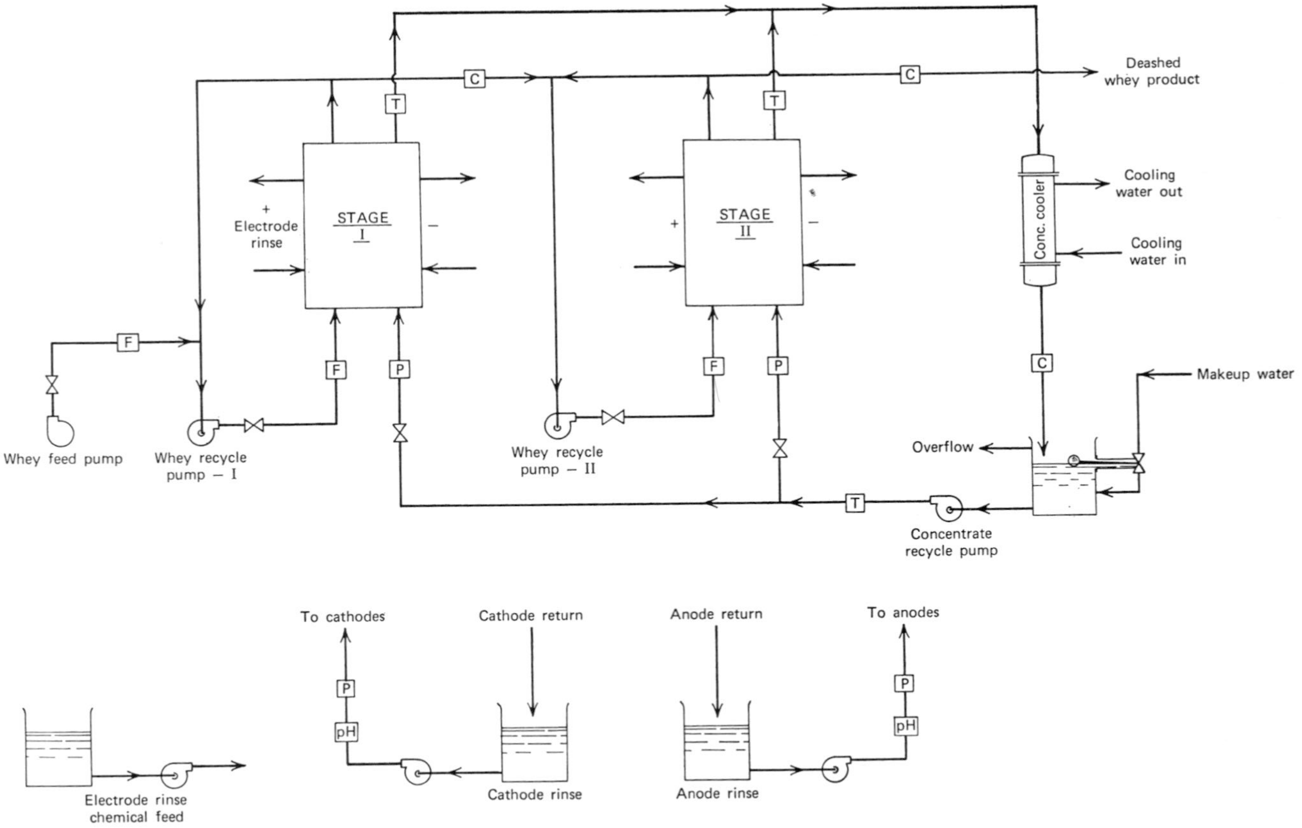

Figure 3. Flow diagram of a whey processing plant. F, flow indicator; P, pressure gage; C, conductivity; pH, pH indicator; T, tem-

66

amortization, is a fixed expense. Figure 4 indicates the approximate order of magnitude of variable operating expenses for treatment of normal whey to

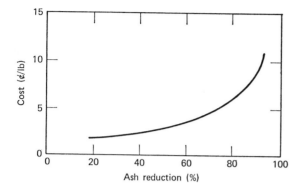

Figure 4. Operating cost versus salt reduction in deashing whey.

various desalting levels. With typical desalting ranges that result in the removal of 50–75% of the ash in whey, these operating costs total less than $0.05/lb of solids.

The basis for these costs must be clearly defined and varies according to specific operating situation. The curve presented is based on an average electrical power rate of $0.01/kW-hr, operating labor at $4.00/hr, membrane life of 1–2 yr, daily chemical cleanup with proprietary dairy cleaning chemicals, and maintenance charges at about 1% per annum of original equipment cost. This curve would be modified slightly depending upon whether transport depletion or normal cation–anion electrodialysis were used. However, the total may be considered about the same since the higher cost of electrical power for transport depletion is offset by the more frequent membrane replacement required for normal electrodialysis. Equipment amortization depends upon the specific accounting procedures for any given operation, but this added burden is approximately equal to the direct variable operating expense.

Figure 5 is a plot of approximate electrical power consumption vs. percent ash removal. In desalting levels varying between 50 and 75%, the electrical power is normally ≤ 0.5 kW-hr/lb of whey solids processed. Once again, this curve varies according to whether transport depletion or normal cation–anion electrodialysis is used. However, the net power consumption in either case is approximately the same. The inherently lower current efficiency of the transport-depletion process is more consistently achieved, whereas the

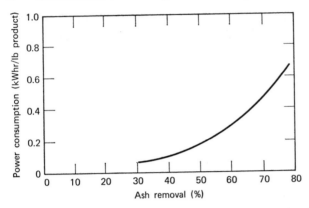

Figure 5. Approximate power consumption for deashing whey (8% ash in feed liquor).

initial high efficiency of normal electrodialysis is offset by severe drops in current efficiency because of the fouling of membranes.

The relationship between percent ash removal and electrical energy consumed per pound of product produced is relatively constant whether the whey is processed at the normal concentration of about 6% solids or processed at higher concentrations of approximately 30–40% solids. Some of the studies of the electrical energy consumption as a function of the concentration of solids in whey indicate that the lowest cost of electrical energy is achieved when the whey contains from 20 to 40% of solids.

The mechanical equipment involved in electromembrane whey-processing systems closely resembles that used in electrodialysis water-demineralization plants. The primary difference in these plants is the use of sanitary stainless-steel-piping systems for whey handling. Food-plant operation under sanitary standards requires the use of piping and pumping systems that can be cleaned by either cleaning-in-place techniques or by system disassembly and steam sterilization. Normal dairy and food-processing operation dictates daily cleaning of the product-handling system.

As discussed previously, electrodialysis stacks themselves are not inherently easy to keep in the sanitary condition normally associated with the processing of food materials. However, careful sanitation techniques and periodic stack disassembly with individual scrubbing of stack parts and membranes normally suffice in keeping stacks in a satisfactorily clean condition.

Other system components such as brine-collecting streams, electrode-rinse systems, conductivity and pH controls, and other auxiliary components are of the same construction that would be used in normal water-desalination plants.

Figure 6 shows the estimated capital equipment cost for whey deashing plants of intermediate sizes. As would be anticipated, equipment for a plant

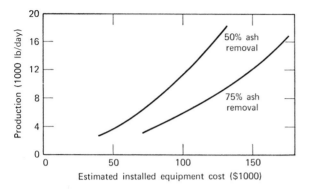

Figure 6. Estimated installed equipment costs for deashing whey (8% ash in feed liquor).

designed to remove a higher percentage of ash would cost more. Removal of more than 85% of the ash results in increased energy requirements and greatly decreased production rates per unit of membrane area. Therefore, operating costs for plants removing more than 85% of the ash are higher than for plants operating at the 50–80% removal level.

Considering all costs of operation, both variable and capital amortization, the cost of whey demineralization typically is in the range of $0.05–0.10/lb of dry solids. This cost level is quite attractive when viewed against the increased value of the demineralized whey solids. In their natural salty form, whey solids are usable primarily for animal consumption at values less than $0.10/lb. In the partially demineralized form, whey is a very nutritious and economical source of protein and finds useful application in many forms of dairy products and synthetic foods with a value in excess of $0.20/lb.

Chapter V   Electromembrane Processes for Recovery of Constituents from Pulping Liquors

Richard M. Ahlgren*

Paper, paper products, and other cellulose-based materials comprise one of the commodities consumed in greatest quantities (other then food materials) in our modern life. It is estimated that annually in the United States almost 100 billion pounds of paper and allied paper products are consumed. This is a per capita annual consumption of almost $1\frac{1}{2}$ lb/day. The United States leads the world in the production of paper goods. It produces about 60% of the total. There are approximately 800 paper mills in the United States turning out various forms of paper and paper products, with about 250 actually processing pulp. The daily output of these mills is about 150,000 tons of paper products.

Paper and paper products are basically forms of cellulosic fiber. The principal source of cellulosic fiber is wood, but many other sources are also used. Among these are such materials as grass, sugar-cane stalk, rags, waste paper, and others.

The basic structure of a tree, in very simple terms, is a bundle of oriented cellulosic fibers glued together with lignin binders covered on the outside by a protective bark or skin. The process of pulping consists of first stripping away the outer bark or skin covering and then, by either mechanical or chemical means, breaking down and separating the cellulosic fibers from their basic chemical binder. The character of the finished fibers depends greatly

* Aqua-Chem, Inc., Waukesha, Wisconsin.

upon the mechanical and chemical conditions used in dividing and digesting the wood.

Three general types of pulping are commonly used. These are mechanical or ground-wood, kraft or alkaline, and sulfite pulping. The first process, mechanical or ground wood, is, as the name implies, the mechanical mastication of wood to subdivide the fibers individually into relatively short lengths (1–3 mm). Ground-wood pulp is relatively economical to produce, but does not have the finest texture or the best physical properties. Perhaps the most common single use of ground-wood pulp is newsprint, of which about 80% is made of mechanically ground coniferous or soft woods. One of the big advantages of mechanical pulping processes is that chemicals are not required, and thus waste solutions of chemicals are not generated.

One of the most common pulping types used in the world today is referred to as the kraft or alkaline pulping method. In this scheme, debarked wood chips are put into a digestor with relatively high-pH sodium solutions for varying lengths of time under predetermined conditions of temperature and pressure. The result of this "cook" is that the alkaline solutions dissolve the lignin binders of the wood and the cellulosic fibers themselves are released. In the digestor blow pits, the pulp is separated from the spent or black liquor, which is in turn sent to the recovery or disposal plant. Various stages of pulp washing and perhaps bleaching are used, and some of these effluents are added to the strong liquor collected from the blow pits. The result is that an organic liquor with approximately 9–15% solids content is produced. For each pound of wood that is charged to the digestor, approximately one-half pound comes out as the recovered cellulosic fibers, and the other one-half pound is organic solids in the spent liquor. These black liquor solids are composed of approximately 60% lignins and 40% carbohydrates (wood sugars) or their derivatives. In the kraft process, the collected black liquor is normally sent to concentrating evaporators where approximately 80% of the water is removed. The remaining concentrate is then charged to special black liquor furnaces or boilers in which the organic matter is burned for its fuel value and an inorganic ash or smelt is collected from the bottom of the furnace for recovery and reuse in the preparation of new liquor. The kraft pulping process basically is designed around a built-in liquor destruction and chemical recovery scheme.

The third pulping process commonly used is referred to as sulfite or acid pulping. The mechanics of sulfite pulping are similar to those described for kraft pulping in that debarked and washed wood chips are placed into a digestor with a chemical dissolving solution. In contrast to the kraft process, in sulfite pulping this dissolving solution is a neutral or acidic form of the sulfite ion. Generally, four different base ions are used for the preparation of various sulfite liquors. These are calcium, sodium, ammonium, and magnesium ions. Varying amounts of the bisulfite ion and free sulfur dioxide are used so that the pH values of this solution may vary from 7 down to about 2. This spent sulfite liquor is collected at solution strengths of anywhere from 8 up to about 14% by weight, depending on the efficiency of liquor-collection systems and the amount of pulp washing performed.

The chemical composition of the liquor varies with the type of wood being used, cooking formulations, and the characteristics of the product being made. Table 1 gives some general chemical compositions of typical kraft

Table 1. Typical Spent Liquor Composition

	Kraft	Ammonia Acid Sulfite	Calcium Acid Sulfite
pH	11.9	3.0	2.2
Total solids (g/l)	150	115	95
Base ion (g/l)	33 (Na)	3 (NH_3N)	4 (Ca)
Lignins (g/l)	75	55	55
Sugars (g/l)	1	30	21
BOD (g/l)	43	37	32
Sulfur dioxide (g/l)		10	6

and sulfite spent liquors. Many proprietary variations are made from these basic formulations. Kraft liquors are used that have various amounts of sodium hydroxide, sodium carbonate, sodium sulfide, and sodium sulfate. In sulfite liquors, the base ion can be calcium, magnesium, sodium, or ammonium, and acidities can vary over wide ranges. When certain types of wood are pulped by the sulfite process, varying amounts of low-molecular-weight acids, such as acetic acid and formic acid, are also found in the spent liquor, in addition to lignins and carbohydrates extracted from the wood.

The application of electromembrane processing to spent liquor fractionation and recovery is relatively new. Research and pilot-scale studies have been initiated within the past ten years. It was quickly recognized that spent pulping liquors present some of the worst conditions for fouling and degradation of ion-selective membranes. Various charged anionic materials in spent liquors would be transported to the anion-exchange membranes and then almost irreversibly deposited on the selective sites of this membrane. Consequently, little or no attempt was made to apply classical cation–anion electrodialysis to the recovery of either kraft liquors or sulfite liquors. In the early 1960's, Dubey, Wiley, and others[1-4] at the Pulp Manufacturers'

Research League in Appleton, Wisconsin, developed a four-stream process for the fractionation and recovery of soluble-base spent sulfite liquors. This initial work concentrated primarily on acid sulfite liquors because an already workable scheme for destruction and recovery of kraft liquor was available. In addition, there were few membranes available at that time that had good resistance to highly alkaline conditions. The bisulfite bases utilized relatively expensive chemicals and thus offered a good chance for satisfactory economics of recovery. Work on by-product utilization also indicated that the various lignin and lignosulfonate forms in spent sulfite liquor probably offered the best chances for utilization as chemical by-products. The research of the Pulp Manufacturers' Research League resulted in the development of the four-stream BALC Process for the electrodialysis of soluble-base sulfite liquors. A diagram of this process is shown in Figure 1.

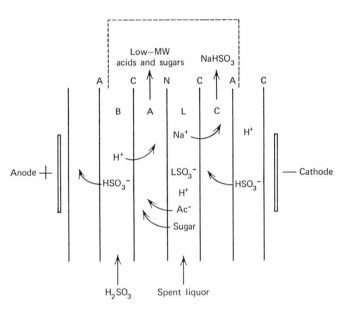

Figure 1. Diagram of the BALC process—four-compartment repeating cell. B, barrier; A, anolyte; L, liquor; C, catholyte; N, neutral membrane.

In this process, the spent liquor at any concentration between 10 and 40% solids is pumped into a chamber bounded on one side (cathode side) by a cation-selective membrane and on the other side (anode side) by a neutral or nonselective membrane. Thus, in contrast to the reverse-osmosis processing of relatively dilute pulp mill wastes that is described in Chapter 11, electromembrane processing is more suitable for treating concentrated pulping wastes. The solutions flowing on either side of the spent liquor compartment are called the anolyte and catholyte solutions. The base ion is recovered as the bisulfite in the catholyte solution. The various organic anions are recovered as acids in the anolyte solutions. Sulfur dioxide is injected into water to form sulfurous acid, and this solution is circulated through the fourth compartment of a four-compartment repeating cell. This barrier stream of sulfurous acid contributes hydrogen ions to the anolyte compartment and bisulfite ion to the catholyte compartment to maintain the electrochemical balance of the system. These repeating four-cell units (referred to as BALC cells) consist of two cation membranes, one anion membrane, and one neutral membrane. Many such repeating units are assembled between end frames that contain electrodes to form a BALC electrodialysis stack.

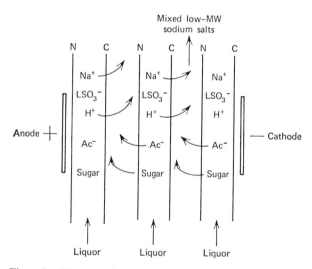

Figure 2. Diagram of the transport-depletion process. N, neutral membrane; C, cation-exchange membrane.

Figure 2 is a diagram of a cation-neutral membrane transport depletion process for recovery of pulping chemicals. This process can be considered as an alternative method of treatment of spent pulping liquors, in that it uses the neutral-membrane technique that avoids fouling of the anion membranes by organic anions. However, a basic problem that exists with the application of either the transport-depletion system or conventional cation–anion membrane electrodialysis to recovery of pulping chemicals is that the pulping base is not recovered in a chemical form that is easily and readily reusable in the digestion plant. In other words, if a sodium-base spent sulfite liquor were electrodialyzed by conventional equipment, the concentrate or ion-enriched streams would be a mixed solution of sodium salts and low-molecular-weight organics. This particular chemical composition is not desirable for use as a digestor liquor. On the other hand, if the goal of the process is to recover organics that are free of cations, this process might be practical.

Other electromembrane methods and techniques have been proposed for treatment of spent pulping liquors. Some of these methods have been designed specifically for use with sulfite liquors, whereas others are intended for use with kraft liquors. One of these processes is an all-cation membrane sodium–hydrogen ion-exchange scheme proposed by Mintz, Lacey, and Lang,[5] as illustrated in Figure 3. In this process, the spent sulfite liquor is run in parallel

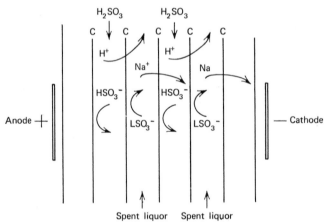

Figure 3.   Diagram of all-cation membrane electrogravitation process. C, cation-exchange membrane.

alternate streams with compartments filled with sulfurous acid (H_2SO_3). The proper selection of cation membranes allows the passage of sodium ion from the spent sulfite liquor, forming the desired sodium bisulfite pulping liquor. The hydrogen ion is depleted from the sulfurous acid stream and is transported into the sulfite liquor, forming lignosulfonic acids. This process would probably function best when operated as an electrogravitational scheme, allowing the denser sodium bisulfite to settle to the bottom of the cells and be drawn off from alternate compartments. However, if cation membranes are used with sufficiently different properties for the transport of hydrogen and sodium ion, the process may be operated as a conventional flowing-stream stack.

A three-compartment electrogravitational method for the fractionation and recovery of kraft black liquor is shown in Figure 4. In this procedure, black

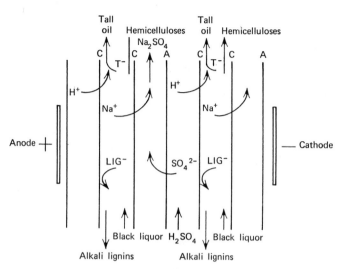

Figure 4. Diagram of the three-compartment electrogravitation process. C, cation-exchange membrane; A, anion-exchange membrane.

liquor is fed into every third compartment of a cation–cation–anion repetitive-unit membrane stack. Sulfuric acid is used as a depleting stream contributing sulfate ions to the sodium ion from the black liquor, forming sodium sulfate and contributing hydrogen ion to the black liquor stream, which could be taken off as a tall-oil by-product. This mechanism proposes a very delicate balance of streams and would have to be operated as an electrogravitational process in order to achieve the desired fractionations.

Several additional arrangements of cation, anion, and neutral membranes have been proposed for use as electromembrane schemes in recovering spent paper pulping liquors. Some of these operate along conventional electrodialysis lines, whereas others are variations of the electrogravitational or electrosorption techniques.

Considerable laboratory and pilot-scale work with electromembrane processing and spent sulfite liquors has been carried out by paper-mill research groups, individual pulp and paper manufacturers, and some educational institutions. Up to the present time, none of these methods has been actually put into routine production operation for the recovery of chemicals or by-product values from spent pulping liquors. Two factors contribute to the lack of industrial application of the process at this time. The first of these is that the economics of pulping base recovery is marginal compared to the purchase of fresh chemical. The second consideration is that broad-scale uses of electrodialyzed spent liquor organic solids have not yet been developed.

With respect to the application of organic liquor solids to other industrial uses, markets have been investigated but not thoroughly developed. One of the primary applications of spent liquor solids is in binder and adhesive uses. Decationized soluble-base sulfite liquors can be polymerized under proper conditions of time, temperature, and pressure into a high-quality organic binder for several products. The primary and most widely investigated use of this type is application of decationized ammonia-base sulfite liquors as a glue or binder in plywood manufacturing operations. Considerable development work has been carried out with this specific product, and plywood samples made with electrodialyzed ammonia-base liquor have been shown to be as good as, or better than, samples made with standard phenolic glues for interior product grades of plywood. Exterior or marine plywood grades utilizing electrodialyzed spent sulfite liquors as binders have not yet been fully developed or evaluated.

Several additional uses are currently being made for spent sulfite liquors. The modified or improved properties exhibited by electrodialyzed liquor for these applications is not yet fully known. Some of the uses for which spent liquors are now utilized are as pelletizing and briqueting binders for taconite iron ore, binders for combustible briquets, edible binders for pelletized animal foods, dispersants, and fluidizing aids.

Application of electrodialysis equipment to spent pulping liquors involves several markedly different conditions from those normally encountered with the electrodialysis of saline waters or even other chemical or biological solutions. Some of these various considerations are as follows:

1. Practical electrodialytic treatment of spent pulping liquors involves operation at high temperatures (approximately 175°F).

2. Electrodialysis of spent pulping liquors involves high concentrations of ionized inorganics (1–4%) dissolved in high concentrations of soluble organics (10% or higher).

3. During electrodialysis of spent sulfite liquors, a peculiar situation occurs in which product-solution conductivity increases with percent demineralization. This phenomenon is the opposite of that usually encountered with electrodialysis, in which solution conductivity decreases as ions are removed.

4. The presence of short cellulosic fibers presents a potential physical fouling situation not present in most membrane processing. These fibers are

the result of normal pulp washing and separation techniques in which fiber bypassing occurs. The use of almost "foolproof" filtration of fibers would be prohibitively expensive.

The ampere-hours required per equivalent of chemical recovered is not a constant value as in normal water desalting. This current requirement decreases rapidly as liquor concentration increases because of the effects of dialysis (in contrast to electrodialysis) on base ion removal. Optimum concentration is a complex function of liquor type and chemistry, but it is usually observed that the requirements for electrical current are less at concentrations between 25 and 40%. Excessively high concentrations (about 50%) present problems with high fluid viscosity and short circuiting of electrical current.

Power requirements per pound of base ion removed are a function of percent removal as well as operating current density. Typical figures for the electrodialyzing of ammonia-base sulfite liquor by the BALC process are shown in Table 2. At a current density of approximately 100 mA/cm², the

Table 2. Power Consumption vs. Ammonia Removal

Percentage NH_3 Removed	At 75 mA/cm² (kW-hr/lb NH_3)	At 105 mA/cm² (kW-hr/lb NH_3)
10	0.5	0.8
20	1.3	2.0
30	2.5	4.0
40	3.0	5.0
50	4.0	6.0

energy requirement is about 6 kW-hr/lb of ammonia removed, when the liquor is depleted about 50%. In contrast to this, only 4 kW-hr is required if the process is performed at a current density of 75 mA/cm².

Some of the results obtained in the research on the BALC process performed at the Pulp Manufacturers' Research League are described in Table 3. Most operations were carried out to remove the pulping bases. Recoveries ranged from 60 to 80%. The current efficiencies observed in these operations range from about 50% to more than 100%. The inordinately high current efficiencies resulted because some of the ions were transported by dialysis in addition to the ions transported electrically. Soluble bases such as sodium are frequently removed with current efficiencies on the order of 100% or higher. On the other hand, ammonia-base sulfite liquors are processed at considerably lower current efficiencies. It is generally thought that more of

Table 3.   Percent Base Removal vs. Current
Efficiency

Base Ion	% Base Removed	% Current Efficiency
Ammonia acid	68	52
Sodium acid	70	90
Sodium neutral	58	108
Calcium acid	77	74
Magnesium acid	95	70

the ammonia in ammonia-base liquors is organically bound and not available for electromembrane removal than is observed with other bases.

Sharp drops in the pH of spent sulfite liquors occur during electrodialysis. This decrease in pH is related to acid formation from the splitting of various salts within the liquor. Also, in BALC cell processing, the addition of hydrogen ions to the liquor from the anolyte stream occurs, forming acid analogs of some of the liquor components.

Estimates of capital equipment required for various quantities of two different pulping liquors are shown in Figure 5. This plot is based on 70% base recovery and average quantities of liquor generated per ton of pulp produced. Divalent pulping bases such as magnesium are more efficiently

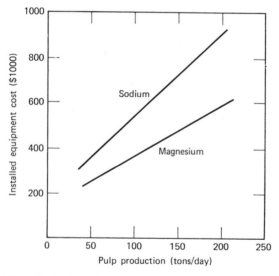

Figure 5.   Estimated equipment cost for 70% base ion recovery.

dialyzed than monovalent solutions such as sodium sulfite liquor. In contrast to this situation, the operating cost of recovery of a monovalent ion such as sodium is less than the cost of magnesium-base recover recovery.

At the present time, the electrodialysis process for recovery of base ions from spent sulfite liquor is of marginally favorable economics. As improved equipment for electrodialysis is developed to operate under higher temperatures and with less stringent pretreatment, this cost situation will become more favorable. In addition uses of spent liquor by-products will increase in the coming years. Besides the expectation of some reduction in net processing costs because of improved equipment and new uses for the by-products, the increasingly stringent restrictions imposed for pollution control will make the process increasingly more attractive. With the development of membranes suitable for high-pH operation, it is anticipated that the electrodialysis process will also become attractive for recovery of organic values from some kraft liquors as well as spent sulfite liquor.

REFERENCES

1. G. A. Dubey, Can. Pat. 677,654 (January 7, 1964).
2. B. Lueck, H. Olsen, and A. J. Wiley, "Modified Spent Sulfite Liquor Products as Binders and Adhesives for Briquettes and Other Products," *Proceedings of The Seventh Biennial Briquetting Conference*, Jackson, Wyoming, August 1961.
3. A. J. Wiley and J. M. Holderby, *Pulp Paper Mag. Can.* **61** (3), T-212 (1960).
4. G. A. Dubey, T. R. McElhinney, and A. J. Wiley, *Tappi* **48** (2), 95 (1965).
5. M. S. Mintz, R. E. Lacey, and E. W. Lang, *Tappi* **50** (3), 137 (1967).

Chapter VI Concentration of Electrolytes Prior to Evaporation with an Electromembrane Process

Tadashi Nishiwaki*

I. INTRODUCTION

A problem often encountered in industrial processing is the concentration of electrolytes in dilute aqueous solutions. In the past, evaporative processes have been preferred to effect such concentrations. However, with the advent of commercially viable processes that use ion-exchange membranes some concentration of dilute aqueous salt solutions can be performed at lower cost than by evaporation alone by combining a pretreatment of electrodialytic concentration with evaporation. The basic principles of the process of electrodialytic concentration are essentially the same as those developed for electrodialytic desalination of saline water, but there are certain problems in electrodialytic concentration of electrolytes that are not encountered in desalination and that require the development of special design features.

The following processes have been used for the concentration of salts in aqueous solution.

1. Evaporative processes are the most commonly used processes for concentrating salt solutions. Single-effect evaporators are characterized by high energy requirements, and multiple-effect evaporators are often used to reduce energy requirements. Even with multiple-effect evaporators, the energy requirements are relatively high, and the cost of amortizing the capital investment required for evaporators, either single- or multiple-effect, is

* Asahi Glass Company, Ltd., Tokyo, Japan.

appreciable. Moreover, if it is desired to separate an electrolyte from a non-electrolyte an additional separation process is needed.

●2. Ion-exchange processes can be used to concentrate electrolytes in solutions. In ion-exchange processes a dilute solution is fed into a column containing ion-exchange resins. Electrolytic components are adsorbed by the resin. The adsorbed electrolytes must then be eluted from the resin by treatment with another solution. Thus, the process requires a two-step procedure. The amount of ion-exchange material required is proportional to the amount of the solute to be recovered. For these reasons the process has usually been found to be uneconomical for mass production. However, the ion-exchange process has been widely used to deionize water, and has found some special uses in industry.

3. With ion-exchange-membrane processes it is possible to recover an ionized solute dissolved in a small amount of water by a continuous process driven by an electrical voltage. With this process simultaneous concentration and separation of electrolytes from nonelectrolytes can easily be accomplished. Although the main industrial use of ion-exchange-membrane processes, so far, has been for recovering electrolytes of relatively low value from dilute aqueous solutions, it would appear, from inquiries received by manufacturers of ion-exchange membranes, that a number of studies are under way in private laboratories for other separations.

Because of unique circumstances a concentration process based on ion-exchange-membrane electrodialysis has been developed in Japan. Japan has no native salt deposits. In addition, the large amounts of rainfall make solar evaporation of seawater a relatively expensive means of producing salt. Nevertheless, prior to the development of electrodialytic concentration processes solar evaporation was the only method used in Japan to produce sodium chloride. The production cost as well as the amount of production depended entirely upon the weather. Because of the high costs and undependable supply of salt in Japan, the salt used as raw material for the alkali industry was all imported.

Alkali manufacturers in Japan, such as the Asahi Glass Company, the Asahi Chemical Company, and the Tokuyama Soda Company, have conducted studies of electrodialytic concentration to produce brines that have almost the same concentration of salt as the brines produced by solar evaporation. In the development of these processes, these companies have had to solve problems that are in some ways more difficult than the problems encountered in desalination. An important result of the development studies is the development of membranes and operating techniques with which only univalent ions, such as sodium and chloride ions, are transferred to the concentrated solutions. Divalent ions, such as calcium ions and sulfate ions, which cause precipitation, are prevented from transferring to the concentrated solutions insofar as possible.

When ion-exchange-membrane electrodialysis is considered for concentrating electrolytic solutions, initial attention should be given to the following points.

- If substances are present that can cause oxidation of the membranes, such as dissolved halogens or nitric acid, they must be removed or rendered inactive.
- The electrolytes to be concentrated should be strongly dissociated, if low-cost concentration is to be effected.
- If substances that will precipitate when concentrated are present in the feed solution (e.g., $CaSO_4$ in seawater) along with the desired solute, they should be removed or other steps (discussed later) should be taken to prevent precipitation.
- If insoluble particulate matter is present in the feed solution, it must be removed by filtration, or other means.

II. DESIRABLE PROPERTIES OF ION-EXCHANGE MEMBRANES USED IN CONCENTRATION PROCESSES

If the highest degrees of electrolytic concentration are to be effected with low energy requirements, the electrolytes must be transferred through the membranes with high coulomb efficiencies and the amounts of water that permeate the membranes and transfer to the concentrated solutions must be small. The relationships between these material transfers and the properties of ion-exchange membranes is discussed below.

The flux of electrolyte to the concentrated solution through a unit area of membrane pair* is

$$\frac{dq}{d\phi} = (\bar{t}_c + \bar{t}_a - 1)\frac{i}{F} - K_s \Delta C \tag{1}$$

where $K_s = \dfrac{D_c}{X_c} + \dfrac{D_a}{X_a}$

$dq/d\phi$ = flux of electrolyte per unit area of membrane pair
$\bar{t}_c$ = transference number of cations through cation-exchange membrane

---

* A membrane pair is one anion-exchange and one cation-exchange membrane, as found in a repeating unit in the electrodialytic concentration process.

i_a = transference number of anions through anion-exchange membrane

i = current density

F = Faraday constant

ΔC = difference in concentration between the dilute solutions and concentrated solutions

D_c = diffusion coefficient of electrolyte through the cation-exchange membranes

D_a = diffusion coefficient of electrolyte through the anion-exchange membrane

X_c = thickness of cation-exchange membrane

X_a = thickness of anion-exchange membrane

The current efficiency is

$$\eta_c = \left[\frac{100F}{i}\right]\left[\frac{dq}{d\phi}\right]$$

or

$$\eta_c = \frac{100F}{i}\{(i_c + i_a - 1) - K_s \Delta C\} \qquad (2)$$

where

$$\eta_c = \text{coulomb efficiency, } \%$$

Equation 2 shows that high coulomb efficiencies are obtained when the transference numbers are high, the diffusion coefficients are low, and thick membranes are used. Because of the decrease in Donnan exclusion in ion-exchange membranes with an increase in electrolyte concentration in the solutions in contact with membranes, the transference numbers of ion-exchange membranes decrease as the degree of concentration increases. In addition, when the degree of concentration is high, the rate of back diffusion is relatively high because of large concentration differences between the dilute and concentrated solutions. In general, higher-quality ion-exchange membranes are needed for concentrating electrolyte solutions than are needed for desalting saline water.

The flux of water transferred to the concentrated solution through a unit area of membrane pair is

$$\frac{dV}{d\phi} = (\beta_c + \beta_a)i + K_w \Delta C \qquad (3)$$

where $K_w = \dfrac{D_{wc}}{X_c} + \dfrac{D_{wa}}{X_a}$

$dV/d\phi$ = flux of water per unit area of membrane pair

β_c = electroosmotic water-transport number through cation-exchange membrane

β_a = electroosmotic water-transport number through anion-exchange membrane

D_{wc} = diffusion coefficient of water through cation-exchange membrane

D_{wa} = diffusion coefficient of water through anion-exchange membrane

By dividing Equation 1 by Equation 3, the concentration of the concentrated solution is obtained.

$$\frac{dq}{dV} = \frac{(t_c + t_a)(i/F) - K_s \Delta C}{(\beta_c + \beta_a)i + K_w \Delta C} \tag{4}$$

Equation 4 shows that if the transference numbers are high, the diffusion coefficients of salt and water are low, the electroosmotic water-transport numbers are low, and thick membranes are used, highly concentrated solutions can be obtained.

It has been found that thick membranes with high densities of ion-exchange groups and low water contents best achieve the foregoing desirable properties of high transference numbers, low diffusion coefficients, and low electroosmotic transport numbers.

To minimize electrical energy requirements for electrodialytic concentrations, the membranes should have the lowest possible electrical resistances. To achieve low electrical resistances, thin membranes with high water contents are desirable. These properties needed for low electrical resistance are opposite to those needed to achieve high coulomb efficiencies and high degrees of concentration. Therefore, in the development of ion-exchange membranes for the electrodialytic concentration process compromises have been necessary.

Before choosing the best membranes for electrodialytic concentration, the relationships between the costs for electrodialytic concentration and the costs for further evaporation should be studied so that the degree of electrodialytic concentration that results in lowest overall production costs can be selected. Once the desirable degree of electrodialytic concentration has been selected, the concentrations of feed solution to the process and brine from the process may be established. Then a choice of the best ion-exchange membranes may be made by measuring transference numbers, electroosmotic water-transport numbers, and diffusion coefficients of salt and water at the two solution concentrations of interest for the membranes of interest, and using the measured properties in Equations 1–4.

III. THE EFFECTS OF THE CONDITIONS OF ELECTRODIALYTIC OPERATION ON THE CONCENTRATION OF THE CONCENTRATED SOLUTION

The conditions of operation that affect the concentration of the concentrated solution are the concentration of the dilute solution, the temperature, and the current density. Examples chosen from studies made on the electrodialytic concentration of seawater are used to illustrate the effects of the above variables on the attainable concentration of brine.

The effect of variations in the concentration of electrolytes in the seawater fed to the process (i.e., the dilute solution) on the concentration of the brine product, as determined by Yamane and co-workers,[1] is shown in Figure 1.

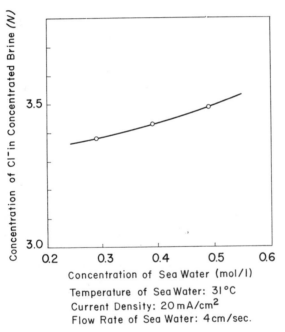

Temperature of Sea Water: 31°C
Current Density: 20 mA/cm^2
Flow Rate of Sea Water: 4 cm/sec.

Figure 1.   Effect of concentration of seawater on the concentration of concentrated brine.

Figure 1 shows that as the concentration of electrolytes in the seawater feed increases (from about 0.25 to about 0.55 mole/l) the concentration of the brine product increases (from about 3.38 to about 3.52 N).

Yamane and co-workers also showed the effect of temperature on the concentration of the brine product, as given in Figure 2. Figure 2 shows that

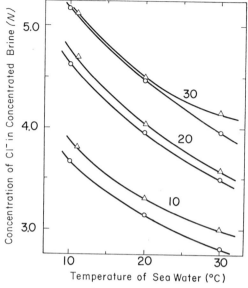

Figures in the graph indicate the current density, m amp./cm^2

Figure 2. Effect of temperature of seawater on the concentration of concentrated brine.

as the temperature of the seawater feed increases (from 10 to 30°C) the concentration of the brine product decreases at all three current densities studied.

The effects of current density on the concentration of product brine, as given by Oda,[2] are shown in Figure 3. With an increase in current density over the range shown (up to 30 mA/cm^2) the concentration of product brine increases, whether measured by the normality of sodium ion present, or by the normality of the total electrolyte present.

IV. PREVENTION OF SCALE FORMATION

When seawater is concentrated by simply removing water (e.g., evaporation), the concentrated solution will become saturated with CaSO$_4$ and precipitates of CaSO$_4$ will form long before the sodium chloride is concentrated to a desirable level. It was thought that if ion-exchange membranes could be

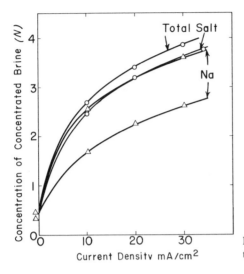

Figure 3. Effect of current density on the concentration of concentrated brine.

developed that would transfer univalent ions (e.g., Na^+ or Cl^-) in preference to divalent ions (e.g., Ca^{2+} or SO_4^{2-}), only univalent ions would be transferred to the concentrated brine solution and precipitation of $CaSO_4$ in the concentrate compartments would be minimized or eliminated. With this use in mind, ion-exchange membranes (both cation-exchange and anion-exchange) were developed that have a higher permeability for univalent ions than for divalent ions. Membranes with such properties are termed univalent-selective ion-exchange membranes, and are made by either (a) preparing homogeneous membranes with resins that have low permeability for divalent ions,[3-7] or (b) coating conventional ion-exchange membranes with a thin film of a resin with low permeability for divalent ions.[8-17] Membranes prepared by the coating method have much lower electrical resistances than those prepared by the homogeneous-membrane process and are therefore preferred for use in the electrodialytic concentration process. Two basic techniques of coating conventional ion-exchange membranes with the thin films have been developed. In one method fairly conventional coating techniques are used, such as spraying, brushing, or doctoring films of polymerizable resins onto the base membranes, followed by initiation of polymerization by thermal or other means. In the other basic method conventional ion-exchange membranes are assembled in an electrodialyzer and univalent selectivity is imparted by introducing a dilute solution of reagents that impart selectivity into the electrodialyzer and passing an electric current through the membranes and solutions to effect electrodeposition of the reagent onto the surfaces of the membranes.

The availability of univalent-selective ion-exchange membranes has made it possible to use electrodialytic methods not only to concentrate solutions but also to separate univalent ions from divalent ions in solution, or to effect both concentration and separation simultaneously.

The specific permselectivity between ions of the same charge (ions A and B), P_B^A, is defined as

$$P_B^A = \frac{t_A C_B}{t_B C_A}$$

where t_A = the transport number of the A ion
t_B = the transport number of the B ion
C_A = concentration of the A ion in the dilute solution
C_B = concentration of the B ion in the dilute solution

In currently available univalent-selective membranes the specific permselectivity of magnesium ions relative to sodium ions, $P_{Na^+}^{Mg^{2+}}$, is the range of 0.2–0.3, that for calcium ions relative to sodium ions, $P_{Na^+}^{Ca^{2+}}$, is 0.3–0.4, and that for sulfate ions relative to chloride ions, $P_{Cl^-}^{SO_4^{2-}}$, is 0.02–0.08. Yawataya[18] has shown that the transport numbers of divalent cations through univalent-selective cation-exchange membranes increase with increases in current density, as shown in Figure 4. These changes in transport numbers cause

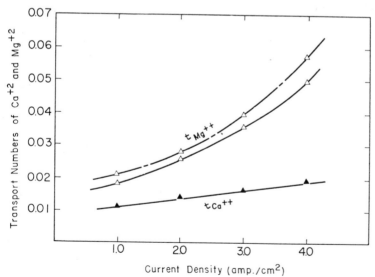

Solid lines: solution 0.45N NaCl + 0.02N CaCl$_2$ + 0.10N MgCl$_2$
Broken lines: solution 0.45N NaCl + 0.10N MgCl$_2$

Figure 4. Effect of current density on transport numbers for Mg^{2+} and Ca^{2+} in univalent-selective membranes.

changes in the specific permselectivities in accordance with the definition of P_B^A given previously. It was also found that the concentration of ions in the dilute solution and the linear velocity of the dilute solution affect P_B^A slightly. Therefore, the current density, the concentrations of ions, and the velocity of the dilute solution must all be considered in selecting the conditions for operation of the electrodialytic concentration process with univalent-selective membranes.

Univalent-selective ion-exchange membranes useful for electrodialytic concentrations and separations are manufactured by the three Japanese firms mentioned previously. These membranes not only have the desirable properties discussed previously, but also are mechanically strong and dimensionally stable. They are stable when exposed to most nonoxidizing chemicals, and have long service lifetimes. Some types of the membranes are homogeneous and some are reinforced with cloth scrims. The cloth-reinforced membranes are stronger, easier to handle, and more dimensionally stable than the unreinforced membranes, but the reinforced membranes usually have slightly higher electrical resistances than the unreinforced ones.

V. ELECTRODIALYTIC CONCENTRATION OF SEAWATER

In Japan, electrodialytic concentration of seawater is performed on an industrial scale to produce sodium-chloride brines, which are further concentrated by evaporation to produce table salt. The details of the electrodialytic process are discussed below to illustrate some of the factors involved in electrodialytic concentration.

A. Source and Pretreatment of Seawater

One of the most important factors in choosing the location of a seawater concentration plant is the assurance of a suitable supply of seawater. The seawater intake point should be located so that the water is not contaminated by land drainage (i.e., runoff from rains), industrial wastes, or municipal sewage wastes. Plankton, which are especially prevalent around the mouth of a river, present serious problems if they enter the electrodialysis stacks. Thus it is best to avoid locating the seawater intake near the mouth of a stream.

The quantity of seawater needed for the production of salt in this process is 100–200 m³(264,000–528,000 gal)/ton of salt. Thus a factory that produces 100,000 tons of salt annually requires about 2000 m³(5,280,000 gal)/hr.

1. Turbidity of the Seawater Feed

Although the annual mean turbidity of coastal seawater is only a few parts per million (ppm), at times the turbidity is as high as 20 ppm, or higher. With proper filtration the turbidity can be reduced to less than 1 ppm during extreme conditions, and to less than 0.5 ppm under normal circumstances.

Pressure-type sand filters, gravity-type sand filters, cascade filters, or precoat-type filters may be used independently or in combination, to remove particulate matter 2μ, or larger, in diameter.

2. Microorganisms

Microorganisms propagate in electrodialysis stacks and pipes and hinder the flow of seawater through the solution compartments and manifolds. Since some types of bacteria can cause deterioration of the ion-exchange membranes, the raw seawater is usually chlorinated to kill the bacteria. However, since solutions containing residual chlorine can cause oxidative attack on the membranes, complete removal, or reduction, of the oxidants in the feed water must be assured before the water enters the electrodialysis stacks.

3. Organic Matter and Other Pollutants

It is not desirable to use seawater with a high content of organic matter as feed water to the process. Seawater containing organic electrolytes, such as synthetic detergents, must be avoided because organic anions cause fouling of anion-exchange membranes. Oxidizing materials in the raw seawater should also be avoided. Organic matter and oxidizing materials can be removed by appropriate treatments (e.g., adsorption with activated carbon, or reduction), but since the pretreatment processes represent added costs, it is advisable to select for feed to the concentration process seawater that does not contain the undesirable materials.

B. Electrodialysis Stacks Used in the Process

The principles of the process for electrodialytic concentration of seawater are the same as those for the process for desalination of saline water. The principles and the general type of multicompartment electrodialysis stacks used are described elsewhere in this book, but the stacks used for the concentration process differ from the stack used for desalination in certain respects. These differing features are discussed below.

Since maximum concentration of brine is desired, usually no solution of any kind is pumped into the compartments from which the concentrated brine is withdrawn. Only the water that enters these compartments by osmosis and electroosmosis is withdrawn along with the ions transferred electrically, as the brine product. In most installations, the concentrated brine is not circulated through the concentrating compartments, but there are a few installations in which brine is recycled through the compartments to minimize the formation of scales and precipitates, and the accumulation of osmotically transferred water is withdrawn along with the ions as brine product.

Industrial-sized electrodialysis stacks may contain several hundred to more than a thousand pairs of membranes (one anion and one cation exchange). The size of the membranes is usually from 1×1 to 1×2 m. Two types of stacks are in industrial use, the filter-press type and the unit-cell type.

Filter-Press Stacks

The filter-press type of electrodialysis stack has been described by Seko.[19] This type of stack is similar to the stacks used for desalination of saline water. Since their features are described elsewhere in this book, only a brief description will be given here. Alternate cation-exchange and anion-exchange membranes are arranged between compartment frames in an assembly similar to a plate-and-frame filter press. The compartment frames are usually about 1 mm thick. Each frame is provided with gaskets at the ends and edges, screens to support the membranes, and manifolding devices in the gasket areas at each end of the frame. The entire assembly of membranes and compartment frames is clamped between end frames that hold the electrodes. Clamping pressure may be applied by bolting or by a hydraulic mechanism at each end.

Unit-Cell Stacks

The unit-cell type of stack has been described in industrial literature by the Asahi Glass Company Ltd.[20] and in a patent by Tokuyama Soda, Ltd.[21] Figures 5 and 6 show some of the details of the unit-cell type of stack. These stacks have a structure especially developed for the concentration of seawater. Each concentrating cell consists of one cation-exchange membrane and one anion-exchange membrane sealed at the edges to form an envelope-like bag. These concentrate cells and screenlike spacers to separate the concentrate cells so seawater can flow between them are alternated to form sections of multicompartments. A number of such sections are placed into rectangular tanks that have electrodes at each end. Concentrated brine overflows each concentrate unit cell through small tubes attached to the unit cells, as shown in Figure 6. More than a thousand unit cells may be assembled into

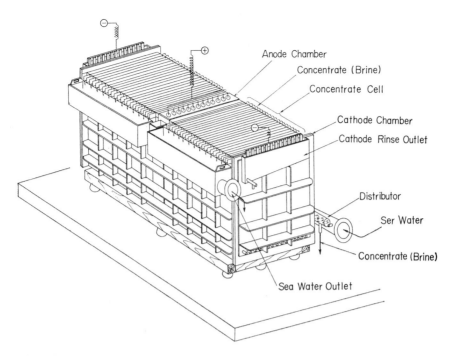

Figure 5.   Unit-cell type of electrodialysis stack.

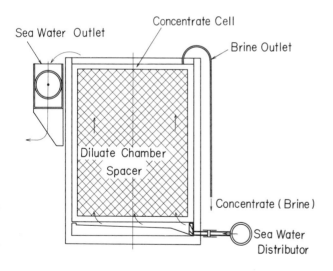

Figure 6.   Main structure of the unit-cell type of electrodialysis stack.

95

each unit-cell stack. A central anode is used with cathodes at each end of the stack, as shown in Figure 5, to keep the total applied voltage to a tolerably low level.

In unit-cell stacks it is just as important to minimize concentration polarization in the dialyzate compartments as it is in filter-press stacks, or desalination stacks. (If severe concentration polarization occurs in the dialyzate compartments, pH changes occur in the concentrate compartments which cause precipitation of pH-sensitive salts, which will plug the compartments.) In unit-cell stacks the seawater is fed to each multicompartment section separately. The flow rates to each section are controlled externally. Inlet-water distributors are provided in each section that are designed to distribute equal amounts of solution to each dialyzate compartment and to distribute the solution equally from edge to edge of the olution compartments. Since concentration polarization is a function of solution velocity, it is extremely important to achieve equal solution velocities in each solution compartment, and to avoid localized stagnation at any location within a compartment. The inlet-water distributors that have been developed adequately maintain equal solution velocities and avoid localized stagnation, and do so without large hydraulic pressure drops. For example, in unit-cell stacks with a height of 40–60 in. (1–1.5 m), superficial linear velocities of 3–5 cm/sec are achieved with hydraulic pressure drops of about 0.7 psi (0.05 kg/cm²). Special net-mesh spacers, as shown in Figure 7, are used in the dialyzate compartments

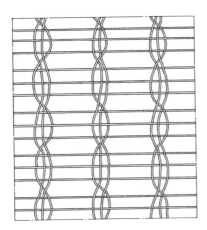

Figure 7. Specially woven spacer net used in the filter-press type of electrodialysis stack.

between the unit cells to support and space the membranes, to aid in the distribution of solutions from edge to edge of the compartments, and to aid in minimizing the thickness of the boundary layers at the surfaces of the membranes (and thus minimize concentration polarization).

To achieve high coulomb efficiencies in the electrodialytic concentration process, it is important to minimize the current through the seawater in the inlet-water distributors and the seawater outlets since current through these two paths bypasses the unit cells and contribute nothing to concentration. Two high-resistance inlet-water distributers are shown in Figure 8. Wilson[23]

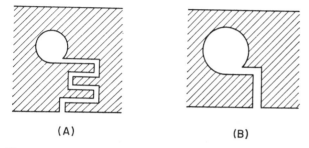

(A)   (B)

Figure 8. Liquor flow paths of cell frames used in the filter-press type of electrodialysis stack.

has an excellent general discussion of electrical leakage current that applies both to the filter-press stacks and the unit-cell stacks.

In unit-cell stacks electrical leakage current is minimized by feeding solution to each multicompartment section through separate inlet-water distributors that are connected to the external seawater distributor through high-resistance (i.e., small-diameter) hydraulic connections. In this way, each multicompartment section is connected to the neighboring sections only by high-resistance paths. The leakage current in unit-cell stacks is usually 1–2%, or less.

Precipitates of pH-sensitive salts in the cathode-rinse compartment that occur as a result of hydroxide ions produced at the cathode can present problems if seawater is used as cathode-rinse solution. It is possible to use special cathode-rinse solutions, but it is generally more economical to provide design features that minimize the problems stemming from the precipitates than it is to use special solutions. In unit-cell stacks the cathode-rinse compartments are designed to provide high solution velocities that sweep away precipitates. The outlet of the cathode-rinse compartment is designed so that precipitates will not deposit on it and block the solution passage, as shown in

Figure 9. In addition, the volumetric flow rate of cathode rinse solution is maintained high enough so the pH of the solution at the outlet never exceeds a value of 9.6, which has been found by experience to provide troublefree operation.

Some of the seawater concentration plants in Japan produce more than 100,000 tons of salt annually. For such plants each electrodialytic stack may produce 10,000 tons/yr, or more. A flow sheet for a seawater concentration

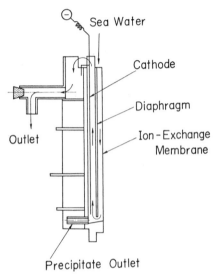

Figure 9.   Cross section of cathode chamber.

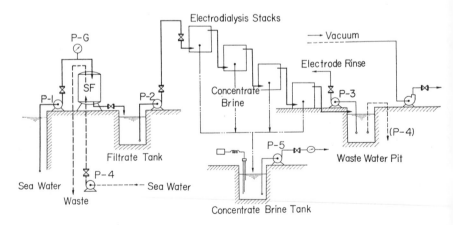

Figure 10.   Flow sheet of seawater concentration plant.

plant is shown in Figure 10. Accordingly, the stacks used have an effective area of ion-exchange membranes 1500–2000 m² (16,000–21,500 ft²). During normal operation it is necessary to inspect each membrane only once or twice during its service lifetime of 3–5 yr.

Some typical design features of filter-press and unit-cell electrodialysis stacks used to concentrate seawater, are shown in Table 1. The typical unit-cell stack has fewer membrane pairs than the filter-press stack, but utilizes a higher percentage of the total membrane area.

Table 1. Specifications of Electrodialysis Stacks

Type	Unit-Cell Type	Filter-Press Type
Size of membrane (cm)	112 × 98	111.5 × 111.5
Effective area of membrane (m²/cell pair)	0.956	0.978
Thickness of dialyzate chamber (mm)	1.4	0.75
Number of membranes (pairs/stack)	1080	1500
Linear velocity of seawater (cm/sec)	3–5	5
Current density (A/dm²)	2–3	3.65

C. Measures Used to Minimize Concentration Polarization

Even though the ion-exchange membranes used for seawater concentration are selective for univalent ions, some calcium ions permeate the cation-exchange membranes and bicarbonate ions permeate the anion-exchange membranes. If severe concentration polarization occurs, hydroxide ions (from ionization of the water) also permeate the anion-exchange membranes and increase the pH of the concentrating compartments. At high pH values calcium carbonate precipitates and blocks the concentrating compartments. In either the filter-press or unit-cell electrodialysis stacks the superficial solution velocities through the mesh-filled dialyzate compartments are maintained at 3–5 cm/sec, which in most cases minimizes severe concentration polarization enough so that current densities of 20–35 mA/cm² can be used. If troubles are encountered with precipitation in the concentrating compartments, the seawater feed is acidified so that hydrogen ions are transferred through the cation-exchange membranes and acidify the concentrating compartments. This acidification converts the insoluble carbonates to soluble bicarbonates.

D. Typical Compositions of Brines from Electrodialytic Concentration

In Table 2 data given by Mitsumi,[24] Onoue,[25] and Kato[26] on the compositions of brines produced at four electrodialytic concentration plants

Table 2.   Compositions of Several Kind of Brines

Ion (N)	Seawater	Brine (Solar Salt Field)	A[a]	B[a]	C[a]	D[a]
			Brine (Ion-Exchange Membrane Electrodialysis)			
Na	0.443	2.335	3.030	2.883	3.170	3.220
K	0.0095	0.050	0.105	0.097	0.150	0.090
Ca	0.0198	0.082	0.075	0.104	0.080	0.100
Mg	0.108	0.553	0.199	0.304	0.120	0.260
Cl	0.523	2.745	3.402	3.366	3.500	3.658
SO$_4$	0.0544	0.275	0.008	0.022	0.020	0.014

[a] A, B, C, and D represent the results from four different electrodialytic concentration plants.

(designated A, B, C, and D) are compared with that produced by solar evaporation. The concentrations of sodium and chloride ions in electro-dialytically concentrated brines are 20–40% higher than those in solar-evaporated brines. The concentration of sulfate ions in the electrodialytically concentrated brines is much lower than that of solar-evaporated brines and even lower than that in the seawater feed.

Pilot-plant studies of electrodialytic concentration of seawater, conducted in 1957, led to the construction of the first commercial installation in 1961.

Table 3.   Specifications of Seawater Concentration Plant (Annual Production of Solid Salt 100,000 tons)

Seawater intake	2,400 m^3/hr
Temperature of seawater (annual average)	18°C
Electrodialysis stack	10 stacks
Ion-exchange membranes	10,000 pairs
Dimensions of membranes	1 × 2 m
Operational conditions	
Current density	4 A/dm
Temperature of seawater (inlet)	30°C
Flow rate of seawater in the stack	5 cm/sec
Results	
Concentrate liquor	
Concentration of chloride	3.8–4.2 N
Concentration of sodium chloride	200–220 g/l
Current efficiency	85–90%
Energy for electrodialysis	250–300 kW-hr/ton NaCl

This first plant had a capacity equivalent to 10,000 tons of salt/yr. Further plant installations and expansions of facilities have been made, until at the present time there is installed capacity to produce more than 200,000 tons of salt/yr. Plants now being designed have capacities from 100,000 to 300,000 tons/yr. Some operational characteristics of a modern 100,000 ton/yr plant are given in Table 3 and in Figure 11. Such a plant is expected to concentrate

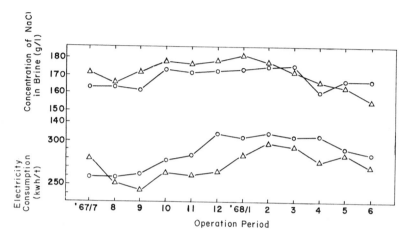

Figure 11. Results of operation of seawater concentration plant.

raw seawater with a 3 % concentration of salt to a concentrated brine (20–22 % salt concentration) with an engery expenditure of about 250–300 kW-hr/ton of salt concentrated.

In large seawater concentrating plants brine can be produced most economically when the electric power facilities are integrated with the salt-concentration facilities, as indicated in Figure 12, to produce electricity for electrodialytic concentration and low-pressure steam for the subsequent evaporation.

The approximate manufacturing costs for concentrated brine produced in such large integrated facilities is shown as Curve A in Figure 13 as a function of annual production rate (tons of salt/yr). The cost of concentrated brine decreases from about $8/ton of salt to about $6/ton as the size of the plant is increased from 50,000 to 300,000 tons/yr. The manufacturing cost for solid salt (including the cost of both electrodialytic concentration and further evaporation) is also shown in Figure 13, as Curve B.

The estimated costs reflect current practice, but cost reductions are expected with further improvements of techniques.

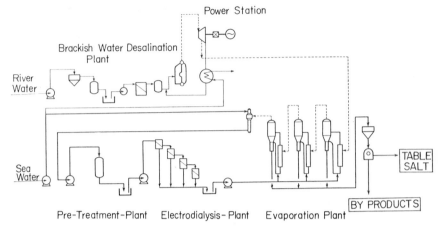

Figure 12. Large-scale salt production plant.

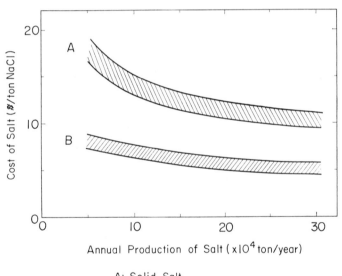

A: Solid Salt
B: Concentrated Brine

Figure 13. Approximate cost of salt in large-scale seawater concentration plant.

VI. OTHER APPLICATIONS OF ELECTRODIALYTIC CONCENTRATION

Several applications of electrodialytic concentration in addition to seawater concentration are now being studied and one or more of them may reach the commercial stage of development. The technique is specially applicable when a desired electrolytic product is to be separated from a mixture of solutes as well as to be concentrated, or when the desired product is heat sensitive or decomposes when heated. Inorganic strong electrolytes, salts of organic acids, and acids and bases can all be concentrated.

Operating results obtained in studies of the electrodialytic concentration of calcium chloride, sodium acetate, and sodium sulfide are given in Table 4.

In the calcium-chloride studies for which data are given in Table 4, the

Table 4. Concentration of Dilute Aqueous Solutions by Means of Ion-Exchange Membrane Electrodialysis

	Calcium Chloride	Sodium Acetate		
A. Salts				
Current density (A/dm^2)	2.0	2.0	4.0	6.0
Concentration of dialyzate (g/l)	29.6	76	74	68
Concentration of concentrate (g/l)	170	188	208	222
Current efficiency (%)	77.0	83.5	81.3	79.3
Energy consumption (kW-hr/ton)	1320	243	496	759
B. Hydrochloric acid				
Current density (A/dm^2)		1.9	5.0	
Concentration of dialyzate				
HCl (g/l)		20.0	4.8	
AlCl$_3$ (g/l)		40.0	3.8	
Concentration of concentrate				
HCl (g/l)		65.0	75.2	
AlCl$_3$ (g/l)		trace	trace	
Current efficiency (hydrochloric acid) (%)		—	14	
Energy consumption (kW-hr/ton of HCl)		650	—	
C. Sodium sulfide				
Current density (A/dm^2)		3.3	4.2	
Concentration of dialyzate (g/l)		27	27	
Concentration of concentrate (g/l)		125	122	
Current efficiency (%)		77.3	78.5	
Energy consumption (kW-hr/ton)		590	790	

feed solution contained dilute CaCl$_2$ (about 2.9 weight %) contaminated by organic impurities. The calcium chloride was not only concentrated to about 16 weight % but separated from the organic impurities.

In the studies of concentrating sodium acetate solutions, the effects of current density on performance were studied. With an increase in current density, the degree of concentration increased, the current efficiency decreased, and the energy requirements increased.

In the sodium-sulfate studies, neither the degree of concentration nor the current efficiency were changed much by a 30% increase in current density. The energy requirement increased approximately in proportion to the increase in current density.

It has been found in other studies that salts of organic acids that have relatively high degrees of disassociation, such as sodium or potassium formates or acetates, can be concentrated effectively by electrodialytic means.

Unlike the concentration of salts, the electrodialytic concentration of acids or bases is usually characterized by low current efficiencies, as is shown in Table 5. Table 5 shows that concentration and separation of HCl from a dilute, impure feed solution are possible, but the current efficiencies are low.

Table 5.   Recoveries of Electroplating Waste Liquors

A. Nickel plating		
Current density (A/dm^2)	1.0	1.5
Waste liquor		
NiSO$_4$ (g/l)	13.63	12.47
NiCl$_2$ (g/l)	1.94	1.81
H$_3$BO$_3$ (g/l)	3.15	3.18
Concentrate		
NiSO$_4$ (g/l)	115.0	133.4
NiCl$_2$ (g/l)	29.9	29.7
Current efficiency[a] (%)	35.8	32.9
B. Copper plating		
Current density (A/dm^2)	1.5	2.0
Waste liquor		
K$_2$Cu(CN)$_3$ (g/l)	8.47	8.41
KCN (g/l)	1.07	1.17
KOH (g/l)	11.67	11.90
Concentrate		
K$_2$Cu(CN)$_3$ (g/l)	91.1	85.6
KCN (g/l)	8.64	9.4
KOH (g/l)	89.1	94.7
Current efficiency[b] (%)	24.7	17.4

[a] Based on nickel ions.
[b] Based on copper ions.

Table 5, which gives operating results from studies made for recovering chemicals from waste liquors from electroplating operations, shows that nickel in waste liquors can be concentrated more than tenfold so that it can be charged back into the plating baths. The copper in waste liquors can be similarly recovered.

SUMMARY

Electrodialytic concentration of seawater to produce a sodium-chloride brine for use in the table salt production has proved to be a commercially viable process in Japan. The equipment, membranes, and operating techniques developed for concentrating seawater almost certainly will be used in the future for other concentrations.

Electrodialytic concentration is specially applicable for products that are heat sensitive or thermally decomposable, and for situations in which a desired electrolytic solute must be separated and concentrated from a mixture of electrolytes and nonelectrolytes.

REFERENCES

1. R. Yamane, T. Sata, and Y. Mizutani, *Bull. Soc. Sea Water Sci. Japan* **20,** 313, 327 (1967).
2. Y. Oda, *Bull, Asahi Glass Res. Lab.* **16** (1), 37 (1966).
3. Tokuyama Soda Ltd., Jap. Pat. 259145.
4. Asahi Chemical Industries, Ltd., Jap. Pat. 270046.
5. R. Yamane, Y. Mizutani, and Y. Onoue, *J. Electrochem. Soc. Japan* **30,** 94 (1962).
6. Y. Onoue, Y. Mizutani, and R. Yamane, *J. Electrochem. Soc. Japan* **29,** 294 (1961).
7. Y. Onoue, Y. Mizutani, R. Yamane, and Y. Takasaki, *J. Electrochem. Soc. Japan* **29,** 187 (1961).
8. Asahi Glass Co., Ltd., Jap. Pat. 281179.
9. Asahi Glass Co., Ltd., Jap. Pat. 282383.
10. Asahi Glass Co., Ltd., Jap. Pat. 291567.
11. Asahi Glass Co., Ltd., Jap. Pat. 410583.
12. Asahi Glass Co., Ltd., U.S. Pat. 3,276,989.
13. Asahi Glass Co., Ltd., U.S. Pat. 3,276,991.
14. Tokuyama Soda, Ltd., Jap. Pat. 236356.
15. R. Yamane, Y. Mizutani, H. Motomura, and R. Izuo, *J. Electrochem. Soc. Japan* **32,** 277 (1964).
16. Y. Onoue, Y. Mizutani, R. Yamane, and Y. Takasaki, *J. Electrochem. Soc. Japan* **29,** 544 (1961).

17. Y. Onoue, Y. Mizutani, W. Teshima, R. Yamane, and S. Akiyama, *J. Electrochem. Soc. Japan* **29,** 468 (1961).

18. T. Yawataya, H. Hani, Y. Oda, and A. Nishihara, *Dechema Monograph.* **47,** 501 (1962).

19. M. Seko, *Dechema Monograph.* **47,** 575 (1962).

20. Asahi Glass Co., Ltd., *SELEMION: Ion-Exchange Membranes,* 1969.

21. Tokuyama Soda Ltd., Jap. Pat. 217865.

22. Y. Tsunoda, *Proceedings of the First International Symposium on Water Desalination, Washington D.C., 1965.*

23. J. R. Wilson, *Demineralization by Electrodialysis,* Butterworth, London, 1960.

24. T. Mitsumi, *Chem. and Chem. Ind.* **21,** 266 (1968).

25. Y. Onoue, *Chemical Plant Engineers Conference, Tokyo, 1968.*

26. M. Kato, *Chemical Plant Engineers Conference, Tokyo, 1968.*

Part II   Pressure-Driven Membrane Processes

Chapter VII   Principles of Reverse Osmosis

Charles E. Reid*

If pure solvent and a solution (or two solutions of different concentrations) are separated by a semipermeable membrane—that is, a membrane which permits solvent, but not solute, to diffuse through it—a flow of solvent through the membrane is commonly observed. If the solvent and solute are under the same pressure, the direction of flow is naturally that which reduces the difference in concentrations, either from the solvent into the solution, or from the more dilute solution into the more concentrated. This flow is the phenomenon called *osmosis*. If the pressure on the solution is increased, osmosis is impeded, and at a sufficiently high pressure it is stopped altogether. This pressure is called the *osmotic pressure*. Still further increase in pressure on the solution reverses the direction of flow, and pure solvent is removed from the solution by passage through the membrane, leaving a more concentrated solution behind. It is this reversed flow that is the basis of the reverse-osmosis method of desalting water.

I. OSMOTIC MEMBRANES

Most studies of osmotic phenomena have depended on one of three types of membranes. The earliest work was done with naturally occurring biological membranes. In 1748, the first known osmotic observations occurred when the Abbé Nollet observed that an animal bladder permitted diffusion of water but not of alcohol. Although osmotic phenomena are essential in the

* Quantum Theory Project, University of Florida, Gainesville, Florida.

mechanism of many biological processes, naturally occurring membranes have not been of much use in either research on osmosis or development of reverse osmosis as a separation process. The first synthetic membrane, made by Traube in 1867, consisted of a gelatinous precipitate formed in the pores of a porcelain plate. This type of membrane, with cupric ferrocyanide as the precipitate, was used extensively from the late 1870's until about 1920. With these membranes accurate measurements of the osmotic pressure of solutions of sugars and a few other organic molecules were made, even up to osmotic pressures of more than 200 atm.[1] A few measurements were made on electrolyte solutions, though these caused rapid deterioration of the membranes. The third type of membrane consists of films of organic polymeric materials. Of these, cellophane films have been widely used in the investigation of polymer solutions, and cellulose acetate membranes have been by far the most widely used membranes for desalination of water by reverse osmosis.

II. THERMODYNAMIC THEORY OF OSMOTIC EQUILIBRIUM

Soon after the publication of Pfeffer's measurements of osmotic pressure in 1877, van't Hoff used them as a basis for building up a theory of solutions. During this same period the development of thermodynamics, largely through the work of J. W. Gibbs, supplied a sound theoretical basis for the study of osmosis and related phenomena of solutions. Through this type of study it was shown that the so-called *colligative properties* are manifestations of essentially the same property of the solution, and that accordingly the same thermodynamic information can be obtained from any of them. Since others were easier to work with than the osmotic pressure, very few measurements of the latter appeared after about 1920. Current interest in osmosis has arisen from its application to the study of polymer solutions and to the development of reverse osmosis as a separation process.

The basic facts needed for a thermodynamic treatment of osmotic equilibrium can be found in practically any textbook on chemical thermodynamics (see, for example, Reference 2). Based on these facts the relation between osmotic pressure and the activity of the solvent can be shown to be[10]

$$\Pi = \frac{RT}{v_1} \ln a_1 \tag{1}$$

where Π is the osmotic pressure,* T is the absolute temperature, v_1 is the partial molar volume, and a_1 is the activity.

* The symbols have the usual meanings attached to them in chemical thermodynamic literature; they are given in the List of Symbols at the end of the chapter.

The above equation is based on the assumption that V_1 is constant, that is, that the solvent is incompressible.

The activity at the vapor pressure of the solution P_1, is often found from

$$a_1 = \frac{P_1}{P_1^*} \tag{2}$$

where P_1^* is the vapor pressure of pure solvent. (If the solute is volatile P_1 must be the partial vapor pressure of the solvent.) This equation is a very good approximation as long as the vapor pressure is low, and can be made exact by replacing the vapor pressures by fugacities. If the usually slight variation of activity with pressure is neglected, this leads to

$$\Pi = \frac{RT}{v_1} \ln \left(\frac{P_1^*}{P_1} \right) \tag{3}$$

When compared with experimental data on sucrose solutions, this equation gives results agreeing within 3% for osmotic pressures ranging up to more than 200 atm.

For dilute solutions $a_1 \approx Z_1$, where Z_1 is the mole fraction of the solvent. If this substitution is made in Equation 2, we have:

$$\Pi v_1 \approx -RT \ln Z_1 = -RT \ln (1 - Z_2) = RT(Z_2 + \tfrac{1}{2}Z_2^2 + \cdots) \tag{4}$$

where Z_2 is the mole fraction of solute. For many solutions, the use of Equation 2 is restricted to solutions so dilute that higher powers of Z_2 may as well be neglected. In this case

$$\Pi v_1 = RTZ_2 \approx \frac{RTn_2}{n_1} \tag{5}$$

or

$$\Pi V_1 = n_2 RT \tag{6}$$

where $V_1 = n_1 v_1$ may be interpreted roughly as the volume of the solvent or as the volume of the solution. Thus under these very restricted conditions, the equation for osmotic pressure closely resembles the ideal gas law, though it is important to note that the resemblance is purely formal. In terms of the molar concentration $C = n_2/V_1$, this becomes

$$\Pi = CRT \tag{7}$$

which is known as the *van't Hoff equation*.

Osmotic pressure is easily connected thermodynamically with the other colligative properties. These all deal with the change in conditions needed to maintain equilibrium between a liquid phase and some other phase (consisting of pure solvent) when the solvent activity in the liquid phase is decreased by the addition of solute. In the treatment of osmotic phenomena the second phase is also liquid, and equilibrium is preserved by changing pressure on the solution. If the second phase is vapor or solid, equilibrium is normally preserved by changing temperature of both phases, and the treatment of boiling-point elevation or freezing-point depression results.

The relation between osmotic pressure and freezing-point depression can be shown to be[10]

$$\Pi = \frac{T \Theta \, \Delta h_{f1}}{V_1 T_f T_f^*} \tag{8}$$

where $\Theta = T_f^* - T_f =$ the freezing-point depression
$T_f^* =$ the freezing temperature of the pure substance
$T_f =$ the freezing temperature of the solution under discussion
$\Delta h_{f1} =$ the heat of fusion of the solvent

Although this equation is not exact, it is often the most practical means of finding osmotic pressures; it is usually much more accurate than calculation from the van't Hoff equation. In fact, the only published data on the osmotic pressure of seawater were calculated by this equation. For seawater with a freezing point of $-1.9°C$, using the values $\Delta h_{f1} = 600.4$ J/mole and $v_1 = 18.05$ cm^3/mole leads to $\Pi = 25.4$ bars (25.1 atm).

If the solute dissociates or associates in solution, all equations connecting osmotic pressure with concentration must be modified, whereas those relating it to solvent activity, vapor pressure, or other colligative properties are still valid. If one formula unit of an electrolyte furnishes v ions on solution, then the osmotic pressure of its solution Π' will be larger than that of a solution of nonelectrolyte of the same molar concentration; the ratio Π'/Π will lie between 1 and v (for not too concentrated solutions) and will approach v as a limit as the concentration approaches zero. For weak electrolytes, the ratio is near one except at extreme dilution, and the reason is primarily that only slight dissociation of such solutes into ions occurs. For strong electrolytes, the ratio is much closer to v and depends strongly on the charges on the ions, polyvalent ions generally giving comparatively low ratios. The reason in this case is that the ions are not fully independent, because of their electrostatic interaction; this is the basis for the presently accepted treatment of dilute electrolytic solutions, the Debye-Hückel theory.

The treatment of osmotic equilibrium given in the preceding section falls short of what is needed to describe the situation in reverse-osmotic flow. The latter involves the essentially irreversible process of water flow through the membrane; moreover, the membranes are not completely impermeable to solute, and so a small solute flow accompanies the water flow. Treatment of such processes as these requires the introduction of nonthermodynamic concepts; however, as discussed by Lonsdale in the following chapter, further insight can be reached through the application of the thermodynamics of irreversible processes. Relevant parts in the general development of this theory are given in the Appendix of this chapter; the fundamental basis can be found in the books by de Groot[3] and Prigogine.[4]

III. MECHANISMS OF MEMBRANE SELECTIVITY

The treatment of osmotic equilibrium by thermodynamics in Section II made no use of any knowledge of the mechanism whereby the membrane selectivity permits solvent, but not solute, to pass through. That this treatment led to an unequivocal expression of the osmotic pressure in terms of the properties of the solution shows that the conditions of osmotic equilibrium are entirely independent of the manner in which the membrane functions. The attempt to go beyond this and treat the nonequilibrium situation in which reverse-osmotic flow occurs, however, requires the introduction of the phenomenological coefficients as discussed in the Appendix. These are nonthermodynamic quantities, and any approach to understanding them must involve those details of membrane mechanism that are irrelevant to the equilibrium situation. For this reason a qualitative description of several mechanisms by which a membrane may work is worthwhile. This description has previously been given by the author[10] and will be summarized here.

If a cellophane membrane is used to measure the osmotic pressure of a solution of a polymer in a solvent of low molecular weight, such as polystyrene in toluene, there is no difficulty in understanding the selectivity of the membrane; it works simply as a *sieve*. This mechanism clearly requires solute molecules much larger than the solvent molecules, with a membrane whose pore diameter is intermediate between them. These requirements appear to rule out this type of mechanism for desalination membranes.

If a solution is separated from solvent by a vapor space, and the solute is nonvolatile, then the vapor space, together with the interfaces, acts as a semipermeable membrane. Desalination of water by a distillation-type osmotic membrane has been carried out on a small scale by Hassler.[7]

If solvent is adsorbed onto the pore walls of the membrane, this adsorbed solvent may so fill the pores as to block passage of solute molecules. Solvent molecules can pass through by moving from one adsorption site to the next. Since the number of adsorbed solvent molecules remains constant, and no one of them need be entirely separated from the adsorbing surface during its migration, the energy requirements for solvent migration are relatively low. On the other hand, solute molecules pass through only by displacing solvent molecules from adsorption sites, with consequent high energy requirement, if they are not also strongly adsorbed. A mechanism of this type appears to be the most probable explanation of the functioning of cellulose acetate as a desalination membrane.[8]

A membrane with fixed charges on its polymer network can act as an osmotic membrane for electrolytic solutes. Membranes that can accomplish osmotic desalination by this means are known.[9]

Finally, it has often been suggested that a membrane may work by dissolving the solvent, the solute being insoluble in the membrane material. It is possible to set up situations in which this mechanism is undoubtedly correct, but in others its meaning is questionable. In a cellophane or cellulose acetate membrane, for example, is water that is adsorbed on the pore walls to be considered dissolved, or must it actually enter the polymer network itself? If solubility of the solvent in the membrane is to be offered as an explanation of semipermeability it is important that such questions as this be faced.

IV. SOME CONSIDERATIONS REGARDING APPLICATION

Applying the principles of reverse osmosis to the actual construction of a separation plant requires, of course, a major development effort. A great deal of such work has been done, and more is currently underway. The aim of this section is not to describe this work, which is covered in subsequent chapters, but to explore the question of how much an understanding of the principles can tell about the requirements for a reverse-osmosis plant.

Separation of water from a solute requires at least a certain minimum amount of energy. This can be found by a thermodynamic calculation, since it is just the amount of energy required when the separation is carried out reversibly. Reverse osmosis approaches a reversible process when the pressure barely exceeds the osmotic pressure. If one mole of solvent passes through the membrane, the volume of solution is decreased by v_1, and so the work done on it is Πv_1. Using the values of Π and v_1 given previously leads to a minimum work for seawater of 25.1×0.018 l-bar/mole $= 45.2$ J/mole, which is about 2.65 kW-hr/1000 gal. Since this is the thermodynamic minimum, it applies to all separation processes for seawater, not only to reverse osmosis.

Like all other processes, reverse osmosis requires in actual use much more energy than the minimum, perhaps several times as much. The most obvious of the several reasons for this is that a pressure just exceeding the osmotic pressure will produce only a very small rate of flow; to achieve a practical rate may require a pressure several times the osmotic pressure. As fresh water is removed through the membrane, the concentration, and hence the osmotic pressure, of the solution left behind is increased. At some point this will necessitate discarding the concentrated solution, either because the osmotic pressure becomes too high or for some secondary reason such as scale deposition on the membrane surface. Since the discarded solution will be at high pressure, recovery of energy from it will be possible, and in large installations it will probably be economically feasible. The efficiency of such energy recovery may strongly influence the optimum amount of water to be removed from the feed before it is discarded.

The increase in concentration as water is removed takes place at the membrane surface, where its undesirable effects are strongest. If no stirring is used, diffusion of solute from the surface back into the bulk solution will occur, resulting in a dynamic equilibrium in which the concentration at the surface will maintain a steady value above that in the solution as a whole. Mechanical stirring can reduce this concentration increment at the membrane surface and so improve the flow rate without added pressure. However, ingenuity is needed to provide this stirring without unduly complicating the apparatus, as will be discussed in later chapters.

Packing the large membrane area required for reverse osmosis into an apparatus of reasonable size is a problem that has attracted much attention. Since the membranes have little mechanical strength but must withstand pressure differences that may reach 100 atm, careful support is necessary, and the support must not seriously interfere with the flow of water. The most obvious way to achieve this is with a unit similar to a plate-and-frame filter press, but units based on tubular or spiral-wound designs are under investigation also. There is one design that obviates the need for mechanical support; the membrane consists of tubes not more than 100 μ in diameter, with the solution on the inside. Because of the small diameter, the membrane can withstand without help the force exerted by the high-pressure solution.

V. CONCLUSION

When the author proposed in 1953 that the Office of Saline Water support a research project on reverse osmosis for desalination, the reasons given were as follows: (a) the process appeared capable of approaching closer in practice to a thermodynamically reversible process than others, and so should have a lower energy requirement; (b) there is no need of high temperatures, with the attendant difficulties such as corrosion; and (c) since reverse osmosis as an industrial process was a neglected field, rapid advance seemed likely once a major development effort was started. Objections raised included doubt that a suitable membrane would ever be found, and a belief that the process would remain only an impractical laboratory curiosity in any case. Fortunately, the Office of Saline Water chose to support the proposed work, and in the course of this project the usefulness of cellulose acetate as a desalination membrane was discovered. This early work has been extended enormously at other research centers, and the potential value of the process now appears well on its way to realization. Now, as discussed in succeeding chapters it will be seen that reverse osmosis is being used and considered for use in industrial applications also.

APPENDIX. DEVELOPMENT OF FUNDAMENTAL FLOW EQUATIONS BY IRREVERSIBLE THERMODYNAMICS

Suppose that the state of the system is characterized by a set of thermodynamic variables $\xi_1, \xi_2, \ldots$, which have the values $\xi_1{}^0, \xi_2{}^0, \ldots$ when the system is at equilibrium. We can now express the entropy relative to the equilibrium state, ΔS, by an expansion in terms of the deviations $\alpha_i \equiv \xi_i - \xi_i{}^0$ of the ξ's from their equilibrium values. Since the value of the entropy is a maximum at equilibrium, linear terms in this expansion must vanish. Usually the treatment is confined to deviations small enough to justify neglecting cubic and higher terms. With this restriction the expansion becomes a quadratic form:

$$\Delta S = -\tfrac{1}{2} \sum_{i,j} g_{ij}\alpha_i\alpha_j = -\tfrac{1}{2}\tilde{\alpha}\mathbf{g}\alpha$$

where the g's are coefficients; in the second form α is a column vector whose coefficients are the α's, the tilde indicates a transpose, and $\mathbf{g}$ is the matrix with elements g_{ij}. Since ΔS must be negative, $\mathbf{g}$ must be a positive definite matrix. Now the *fluxes* (also called *flows* or *currents*) J_i are the time derivatives of the state variables α. Again using vector terminology, we have

$$\mathbf{J} = \dot{\alpha}$$

the dot indicating differentiation with respect to time.
   We further describe as *forces* the elements of the vector

$$\mathbf{F} = -\mathbf{g}\alpha$$

the force F_i being conjugate to the flux J_i. The irreversible process produces entropy at the rate

$$\dot{\Delta S} = -\tfrac{1}{2}(\dot{\tilde{\alpha}}\mathbf{g}\alpha + \tilde{\alpha}\mathbf{g}\dot{\alpha}) = -\dot{\tilde{\alpha}}\mathbf{g}\alpha = \widetilde{\mathbf{J}}\mathbf{F} = \widetilde{\mathbf{F}}\mathbf{J} \qquad (9)$$

since **g** can always be chosen to be symmetric. This shows that the forces and fluxes determine the entropy production through purely thermodynamic considerations. To express the manner by which the forces determine the fluxes, however, requires the introduction of kinetic coefficients which express the susceptibility of the system to the influence of the forces. These are called *phenomenological coefficients* and will be designated by L_{ij}, the set of them constituting a matrix **L**. The relationship between the forces and fluxes is then given by

$$\mathbf{J} = \mathbf{LF} \tag{10}$$

Substituting into Equation 7 then gives

$$\mathbf{S} = \widehat{\mathbf{F}}\mathbf{L}\mathbf{F} = \widehat{\mathbf{J}}\mathbf{L}^{-1}\mathbf{J}$$

Since entropy production must be positive for any natural process, **L** is a positive definite matrix (which assures the existence of its inverse $\mathbf{L}^{-1}$). This means that all main diagonal minors of all orders are positive definite. In particular, all diagonal elements L_{ii} are positive, and for all $i \neq j$ the determinant

$$\begin{vmatrix} L_{ii} & L_{ij} \\ L_{ji} & L_{jj} \end{vmatrix} > 0 \tag{11}$$

This brings us to Onsager's theorem, which is fundamental to the entire theory. This theorem states that with the fluxes and forces defined as above, the matrix **L** is symmetric; that is, that $L_{ij} = L_{ji}$. The proof is based on statistical mechanics.

An immediate consequence of Onsager's theorem is that Equation 11 becomes

$$L_{ii}L_{jj} > L_{ij}^2$$

and $L_{ij} = 0$ if either L_{ii} or L_{jj} vanishes. In other terms, if a force does not cause its conjugate flux to occur, neither will it cause any other flux to occur, nor will its conjugate flux be brought about by any other force. With ordinary choices of properties to describe the state of the system, the off-diagonal coefficients L_{ij} are usually positive or zero, though this is not a fundamental requirement of the theory, and negative L_{ij}'s are possible.

The procedure described here does not uniquely fix the forces and fluxes. If the set of forces F is replaced by a linear combination $\mathbf{F}' \equiv \mathbf{BF}$, where B is a nonsingular coefficient matrix, then simple use of matrix algebra converts Equation 10 to

$$\tilde{\mathbf{B}}\mathbf{J} = \tilde{\mathbf{B}}\mathbf{L}\mathbf{B}\mathbf{B}^{-1}\mathbf{F} \tag{12}$$

and if we set $\mathbf{J}' = \tilde{\mathbf{B}}\mathbf{J}$ and $\mathbf{L}' = \tilde{\mathbf{B}}\mathbf{L}\mathbf{B}$, $\mathbf{L}'$ will be symmetric, and the equation will have the same form as Equation 10. It is a well-known theorem in matrix algebra that a real symmetric matrix may be diagonalized by a transformation of the type in Equation 12; this means that it is possible to choose the forces so that all cross-terms vanish, and each force influences no flux except its conjugate flux. This choice, however, is not necessarily the one that appears most natural conceptually.

Now consider a reverse-osmosis apparatus, designating the properties on the high-pressure side by I, those on the low-pressure side by II. If there is only one solvent and one solute, the equation for thermodynamic equilibrium on either side of the membrane becomes

$$T \, dS = dU + p \, dV - \mu_1 \, dn_1 - \mu_2 \, dn_2$$

n being the number of moles. Fluxes from side I to side II cause an entropy change given by

$$d(S^{II} + S^{I}) = \frac{dU^{II}}{T^{II}} + \frac{dU^{I}}{T^{I}} + \frac{p^{II} \, dV^{II}}{T^{II}} + \frac{p^{I} \, dV^{I}}{T^{I}}$$

$$- \frac{\mu_1^{II}}{T^{II}} \, dn_1^{II} - \frac{\mu_1^{I}}{T^{I}} \, dn_1^{I} - \frac{\mu_2^{II}}{T^{II}} \, dn_2^{II} - \frac{\mu_2^{I}}{T^{I}} \, dn_2 \, dn_2^{I} \quad (13)$$

The quantities dU^{I}, dS^{I}, and dS^{II} can be split into internal and external parts, designated by subscript i and e, respectively. The external energy change $d_e U^{I}$ represents the energy that part I of the system receives from its surroundings (excluding part II), and $d_i U^{I}$ represents energy received by part I from part II (if negative, of course, it will represent an energy flow in the other direction). The interpretation of the others is analogous. The conservation of matter and energy then requires that

$$d_i U^{II} = -d_i U^{I} = dU$$
$$dn_1^{II} = -dn_1^{I} = dn_1 \quad\quad (14)$$
$$dn_2^{II} = -dn_2^{I} = dn_2$$

the third symbol in each case being used for convenience. Moreover, the external entropy change is given by

$$d_e S^{I} = \frac{d_e U^{I} + p^1 \, dV^{I}}{T^{I}}$$

with a similar equation for $d_e S^{II}$ Subtracting these from Equation 13 and using Equation 14, we find for the total internal entropy production

$$d_i S = \Delta\left(\frac{1}{T}\right) dU - \Delta\left(\frac{\mu_1}{T}\right) dn_1 - \Delta\left(\frac{\mu_2}{T}\right) dn_2$$

where Δ has been used to indicate (value in part II − value in part I). If converted to time derivatives by division by dt, this takes the form of Equation 9, with

$$J_1 = \frac{dn_1}{dt} \quad F_1 = -\Delta\left(\frac{\mu_1}{T}\right) = (\mu_1 \Delta T - T \Delta\mu_1)/T^2$$

$$J_2 = \frac{dn_1}{dt} \quad F_2 = -\Delta\left(\frac{\mu_2}{T}\right) = (\mu_2 \Delta T - T \Delta\mu_2)/T^2$$

$$J_3 = \frac{d_i U}{dt} \quad F_3 = \Delta\left(\frac{1}{T}\right) = -\frac{\Delta T}{T^2}$$

If the temperature is uniform ($\Delta T = 0$), it is convenient to concentrate on the matter fluxes only, and write the phenomenological equations as

$$J_1 = -\frac{L_{11}(\Delta\mu_1)}{T} - \frac{L_{12}(\Delta\mu_2)}{T}$$

$$J_2 = -\frac{L_{21}(\Delta\mu_1)}{T} - \frac{L_{22}(\Delta\mu_2)}{T}$$

Since T is constant, further simplification can be achieved by setting $l_{ij} = L_{ij}/T$, so that

$$J_1 = -l_{11} \Delta\mu_1 - l_{12} \Delta\mu_2$$
$$J_2 = -l_{21} \Delta\mu_1 - l_{22} \Delta\mu_2$$

with $l_{12} = l_{21}$ by Onsager's theorem. Clearly it is desirable to have the off-diagonal coefficients l_{12} as small as possible; to make the solvent flux J_1 large, $\Delta\mu_1$ must be large, and this will cause a large solute flux unless l_{12} is small. If l_{12} is small, some qualitative idea of membrane performance can be obtained by the following considerations. Neglecting the off-diagonal terms, we have by Equation 3

$$J_1 \approx -l_{11} \Delta\mu_1 = -l_{11}\left[v_1 \Delta p + RT \ln\left(\frac{a_1^{II}}{a_1^{I}}\right)\right]$$

$$J_2 \approx -l_{22} \Delta\mu_2 = -l_{22}\left[v_2 \Delta p + RT \ln\left(\frac{a_2^{II}}{a_2^{I}}\right)\right] \tag{15}$$

To make this more concrete, for a 3% solution of sodium chloride $a_1^{I} = 0.984$ and a_1^{II} will be practically 1. Both v_1 and v_2 are about 18 cm³/mole, and a_2^{I}/a_2^{II} will vary from, say, 10 for a rather poorly rejecting membrane to perhaps 100 for a very good one. If Δp is 100 bars, we calculate for $T = 300°K$

$$-v_1 \Delta p \approx -v_2 \Delta p = 1.8 \text{ l-bar/mole} = 180 \text{ J/mole}$$
$$RT \ln (a_1^{I}/a_1^{II}) = -40 \text{ J/mole}$$
$$RT \ln (a_2^{I}/a_2^{II}) = 5800\text{–}11{,}600 \text{ J/mole}$$

Thus the predominant driving force for water flow is the term $-v_1 \, \Delta p$, which varies linearly with pressure. On the other hand, salt flow is determined primarily by the term $RT \ln (a_2{}^I/a_2{}^{II})$, and the pressure-dependent term is trivial. It follows that increasing pressure difference increases water flow almost proportionately but has little effect on salt flow. This is Clark's[5] explanation of the often-observed fact that salt rejection improves with increasing pressure. Presumably Clark's mechanism accounts for much of this effect, though the mechanical decrease of pore size with increasing pressure must be taken into account also.[6]

It should be emphasized that the coefficients l_{11}, l_{12}, and l_{22} depend on the equilibrium state of the entire system, including the membrane and the solution; in particular, they are dependent on concentration. This can be seen clearly by considering a situation in which no change of concentration occurs on passage through the membrane. The terms involving activities in Equation 15 then vanish, and if $v_1 \approx v_2$, we find

$$\frac{J_1}{J_2} = \frac{l_{11}}{l_{22}}$$

But in this case, the ratio of flow rates (expressed in moles per unit time) is the same as the ratio of mole fractions, Z_1/Z_2. It follows that l_{11} and l_{22} themselves must be proportional to the mole fractions:

$$l_{11} = l'_{11} Z_1{}^I \qquad l_{22} = l'_{22} Z_2{}^I \tag{16}$$

Now to get 99% salt rejection means that we must have

$$\frac{J_2}{J_1} = 0.01 \, \frac{Z_2{}^I}{Z_1{}^I}$$

Assuming that the values in Equation 16 may be used in Equation 15, we find with the help of the numerical results given above

$$\frac{l'_{11}}{l'_{22}} \approx 8 \times 10^3$$

Nothing in either equilibrium thermodynamics or thermodynamics of irreversible processes tells us anything about how to achieve this ratio, or what types of membranes may exhibit it. Study of these questions requires detailed consideration of the molecular structure of the membranes, and there is now no general theory that can offer much beyond purely qualitative understanding.

LIST OF SYMBOLS

a	Activity
B	Coefficients for transforming fluxes
F	Generalized forces in phenomenological expressions
g	Elements of the entropy matrix
G	Gibbs free energy
h	Molar, or partial molar, enthalpy
J	Fluxes
l	$l_{ij} = L_{ij}/T$
L	Phenomenological coefficients
n	Number of moles
p	Pressure
P	Vapor pressure
R	Gas constant
S	Entropy
T	Temperature (absolute)
t	Time
U	Energy, or internal energy
V	Volume
v	Molar, or partial molar, volume
Z	Mole fractions
α	Deviations of properties from equilibrium
θ	Freezing-point depression
μ	Chemical potential
ν	Number of ions in a formula unit of a salt
ξ	General thermodynamic property
Π	Osmotic pressure

Subscripts

1	Solvent
2	Solute
e	External
f_0	Freezing or fusion
i	Internal (in such symbols as d_iS)

Superscripts

I	Upstream phase
II	Downstream phase

*   Designates a property of a pure substance
l   Liquid
s   Solid

REFERENCES

1. H. N. Morse, *The Osmotic Pressure of Aqueous Solutions*, The Carnegie Institute of Washington, Washington, D.C., 1914; Earl of Berkeley and E. G. J. Hartley, *Proc. Roy. Soc. (London) A* **73**, 463 (1903), **78**, 68 (1906); A. Findlay, *Osmotic Pressure*, Longmans, Green, New York, 1913.

2. E. A. Guggenheim, *Thermodynamics—An Advanced Treatment for Chemists and Physicists*, 4th Ed., Interscience, New York, 1959 or other editions; C. E. Reid, *Principles of Chemical Thermodynamics*, Reinhold, New York, 1960.

3. S. R. de Groot, *Thermodynamics of Irreversible Processes*, North-Holland, Amsterdam, or Interscience, New York, 1952.

4. I. Prigogine, *Introduction to the Thermodynamics of Irreversible Processes*, 3rd Ed., Interscience, New York, 1967.

5. W. E. Clark, *Science* **138**, 148 (1962)

6. C. E. Reid and H. G. Spencer, *J. Phys. Chem.* **64**, 1487 (1960).

7. G. L. Hassler and J. W. McCutchan, *Saline Water Conversion* (*Advances in Chemistry* Series, No. 27), American Chemical Society, Washington D.C. 1960, pp. 192ff.

8. C. E. Reid and E. J. Breton, *J. Appl. Polymer Sci.* **1**, 133 (1959); C. E. Reid and J. R. Kuppers, *J. Appl. Polymer Sci.* **2**, 264 (1959).

9. J. G. McKelvey, Jr., K. S. Spiegler, and M. R. J. Wylie, *J. Phys. Chem.* **61**, 174 (1957); C. E. Reid and H. G. Spencer, *J. Appl. Polymer Sci.* **4**, 354 (1960).

10. C. E. Reid, "Principles of Reverse Osmosis," in *Desalination by Reverse Osmosis*, U. Merten, ed., M. I. T. Press, Cambridge, Mass. (1966).

Chapter VIII   Theory and Practice of Reverse Osmosis and Ultrafiltration*

H. K. Lonsdale†

I. INTRODUCTION

Pressure-driven membrane separation processes have been known for over a century, and until the mid-1950's the term ultrafiltration was generally used to describe such processes when applied to liquids. At about that time it was recognized that at pressures in excess of the osmotic pressure, water could be demineralized by passage through a semipermeable membrane, and the term reverse osmosis came into use to describe this application. In this chapter, we attempt to reduce to practical terms the principles of reverse osmosis given in Chapter 7. In addition, typical performance characteristics of reverse-osmosis membranes are presented, along with some of the limitations of these membranes and of the process in general. Ultrafiltration and the performance of ultrafiltration membranes are treated similarly but in less detail because of the smaller volume of modern literature in this field. This chapter is intended as an introduction to the design and cost considerations and the several practical applications of both processes described in subsequent chapters. The discussion is limited to aqueous phases and to the use of hydrostatic pressure as the driving force even though certain membranes useful in reverse osmosis are also useful in gas separations, which are discussed elsewhere in this book.

* Work supported by the U.S. Department of the Interior, Office of Saline Water, Contract 14-30-2609.
† Gulf General Atomic Incorporated, San Diego, California.

There is clearly no great distinction to be made between reverse osmosis and ultrafiltration. Pressure is used in conjunction with semipermeable membranes to effect separations in both cases. In practice, however, the separations achieved and the membranes and operating conditions used are usually quite different. Reverse-osmosis membranes are generally considered to be those that are highly semipermeable with respect to low-molecular-weight solutes, including salts such as sodium chloride. Ultrafiltration membranes have little or no permselectivity for such solutes; in fact, one of their important applications lies in the separation of colloids or macromolecules, which are retained by the membrane, from low-molecular-weight solutes, which pass through. There is obviously an interface between the two processes that must remain somewhat arbitrary. Michaels[1] has suggested that the separation of solutes whose molecular dimensions are within one order of magnitude of that of the solvent be called reverse osmosis, and that the term ultrafiltration be used to describe separations between solutes having dimensions greater than ten solvent diameters but less than the resolution of the optical microscope (about 0.5 μ). An alternate but still arbitrary differentiation is that in reverse osmosis there is usually a significant osmotic pressure to overcome, whereas in ultrafiltration the osmotic pressure difference across the membrane is usually trivial; thus, reverse osmosis is usually carried out at high pressure (typically 50 atm), whereas ultrafiltration is performed at low pressure (typically <5 atm). The use of the term hyperfiltration[2] in place of reverse osmosis has some merit, but the latter is more widely used.

The reverse-osmosis process began to receive widespread attention in the early 1960's. The potential of the process as a new unit operation for separations was recognized early, and its successful reduction to practice in the water-desalination application has led to consideration of numerous other applications as well as a proliferation of designs of reverse-osmosis equipment. Historically, reverse osmosis can be traced to the finding of Reid and his students[3,4] that cellulose acetate membranes exhibit a remarkable selectivity between salt and water, and the discovery by Loeb and Sourirajan[5] of a method for preparing high-flux cellulose acetate membranes that have an extremely thin selective layer backed by a highly porous substrate. Those developments and a detailed description of the status of reverse-osmosis desalination through 1965 have been covered in a book[6] and an extensive book chapter[2] devoted to the subject. Since that time, a number of field-test results have become available and at least two important new types of reverse-osmosis membranes have been developed. Much of the stimulation and support for reverse-osmosis research and development have come from the Office of Saline Water of the U.S. Department of the Interior.

Reverse osmosis and ultrafiltration can be expected to be preferred over other separation processes only under certain circumstances. The application of pressure to a solution increases the thermodynamic activity of each of its components by an amount proportional to the partial molar volume of each component, and thus establishes a driving force for membrane permeation. For many separations of interest these partial molar volumes are not grossly different, and unless the membrane is highly permselective, efficient separation does not occur in a single pass through the membrane. Multistage operation is possible, of course, and has been considered in certain instances. However, if multiple effects are required, other processes such as distillation are usually economically favored. In the processing of food products or other thermally labile materials, however, even multistage membrane separation processes could be favored.

The application of these processes that has received the greatest attention has been water desalination by reverse osmosis. Here the water permeating the membrane is the product of value. In food concentrating, however, the concentrated solution is of value and the permeate is usually of little consequence. In the treatment of municipal or industrial waste waters, both concentrated feed and permeate are important, the concentrate because it has less negative value than the original feed and the permeate for whatever reuse value it might have. In spite of these differences, all these processes are carried out under qualitatively similar conditions and significant concentration of the feed solution is usually desirable.

II. THEORY

There are several theoretical treatments in common use to describe membrane transport of the type considered in this chapter, and three of these are reviewed below. Two involve physical models of the membrane, and the third treats the membrane as a "black box" in the thermodynamic sense.

The earliest treatments of the pressure-driven permeation processes were based on a porous model of the membrane. It is assumed in this model that all flow occurs through pores which comprise a certain fraction of the membrane area and which have a characteristic size distribution. Flow rate and permselectivity are governed by porosity, pore-size distribution, and specific interactions within the pore fluid. This model has obvious conceptual attractiveness, and it is widely used to describe both ultrafiltration and transport through biological membranes.

A second model portrays the membrane as a nonporous diffusion barrier. All molecular species dissolve in the membrane in accordance with phase equilibrium considerations and diffuse through the membrane by the same mechanisms that govern diffusion through solids or liquids. This model is useful for describing the reverse-osmosis process, wherein essentially non-porous, highly permselective membranes are used, but appears to be too simple to describe ultrafiltration membranes.

An alternate theoretical approach lies in the thermodynamics of irreversible processes. Irreversible thermodynamics provides a useful framework within which to consider reverse osmosis, ultrafiltration, and any of the other dissipative processes described in this book, but it is not model-dependent and therefore sheds no light on flow mechanisms.

We begin this section with the solution-diffusion model as applied to reverse osmosis, and then turn to irreversible thermodynamics as applied to the same process. We conclude with the pore model as applied to ultrafiltration.

A. Solution-Diffusion Model

In the solution-diffusion model, each component of the high-pressure solution dissolves in the membrane in accordance with an equilibrium distribution law and diffuses through the membrane in response to the concentration and pressure differences. Thus the flux, J_i, of each component is given by a relationship of the type[7,8]

$$ J_i = -\frac{D_i c_i}{RT} \operatorname{grad} \mu_i = -\frac{D_i c_i}{RT} \left(\frac{\partial \mu_i}{\partial c_i} \operatorname{grad} c_i + v_i \operatorname{grad} p \right) \tag{1} $$

where D_i = the diffusion coefficient of component i within the membrane
c_i = its concentration in the membrane
v_i = the partial molar volume
p = the applied pressure

There are thus two contributions to the driving force for the flow of each component: a concentration gradient and a pressure gradient. Because very large pressures are required to change the chemical potential by an amount equal to the change caused by only a small concentration change, it is generally not practical to operate a reverse-osmosis system in which there exists a large concentration difference of the solvent. For example, to force water through a solute-rejecting membrane from a solution in which the thermodynamic activity of water is 0.9 (e.g., as in a 15% by weight solution of NaCl) requires a pressure in excess of 2000 psi. In practice, therefore, reverse-osmosis processing is limited to solutions not more concentrated in solute than perhaps 1 M. With water denoted as component 1, Equation 1 can be integrated, when the difference in water concentration across the membrane is small, to yield[7]

$$J_1 = -\frac{D_1 c_1 v_1 (\Delta p - \Delta \pi)}{RT \Delta x} \qquad (2)$$

where Δp and $\Delta \pi$ are the differences in applied pressure and osmotic pressure, respectively, across a membrane of thickness Δx. The principal assumptions made in arriving at the integrated form are that the diffusion coefficient is not concentration dependent and that the membrane properties are not pressure dependent. For aqueous solutions and cellulose acetate membranes both assumptions have been tested and found to be essentially valid. Concentration-dependent diffusion coefficients are common, however, in the permeation of organic liquids through polymeric films.[9]

For most cases of interest the solute is highly rejected by the membrane, and for the solute the first term in Equation 1 is much larger than the second.* The integrated form for solute flow, J_2, is then, to a good approximation,

$$J_2 = -D_2 \frac{\Delta c_{2m}}{\Delta x} = -D_2 K \frac{\Delta c_{2s}}{\Delta x} \qquad (3)$$

where c_{2m} = the concentration of solute in the membrane
c_{2s} = the solute concentration in solution
K = the distribution coefficient for the solute
Thus

$$K \equiv \frac{c_{2m}}{c_{2s}} \qquad (4)$$

and it has been assumed here that K is independent of c_{2s}.

When none of the membrane properties is a function of pressure or concentration, the quantity $D_1 c_1 v_1 / RT \Delta x$ can be considered a membrane constant denoted by A, and $D_2 K / \Delta x$ can be treated as a solute permeation constant denoted by B. We thus have

$$J_1 = -A(\Delta p - \Delta \pi) \qquad (5)$$
$$J_2 = -B \Delta c_{2s} \qquad (6)$$

The constants A and B thus provide a complete practical description of membrane performance. The reduction in solute concentration during

* The conditions for which this approximation is valid can be examined by comparing the two terms in integrated form. For an ideal solution, $\partial \mu_2 / \partial c_2 = RT/c_2$ and the two terms are $RT \ln c_2$ and $v_2 \Delta p$. For NaCl, $v_2 = 18$ cm^3/mole; and if the rejection at 50 atm is only 30%, $RT \Delta \ln c_2 = 8700$ cm^3-atm/mole at 25°C and $v_2 \Delta p = 900$ cm^3-atm/mole, and a slight pressure dependence of salt flux could be expected on thermodynamic grounds. However, when the rejection is 98%, $RT \Delta \ln c_2 = 95,000$ cm^3-atm/mole and the pressure term is trivial.

reverse osmosis is usually given in terms of a dimensionless solute rejection, $SR \equiv (c'_{2s} - c''_{2s})/c'_{2s}$, where $'$ and $''$ denote feed and permeate concentrations, respectively, or by a dimensionless solute concentration reduction factor, $CRF \equiv c'_{2s}/c''_{2s}$. The permeate concentration is determined by the relative flows of solute and water:

$$c''_{2s} = \frac{J_2 c''_{1s}}{J_1} \tag{7}$$

and it follows that

$$SR = 1 - \frac{c''_{2s}}{c'_{2s}} = \frac{A(\Delta p - \Delta \pi)}{A(\Delta p - \Delta \pi) + Bc''_{1s}} \tag{8}$$

and

$$CRF = \frac{A(\Delta p - \Delta \pi)}{Bc''_{1s}} + 1 \tag{9}$$

The water concentration in the product solution, c''_{1s}, is essentially 1.0 g/cm³ in the cases of interest here.

It should be noted that in Equations 2–9, the flows of both water and solute are inversely proportional to the membrane thickness but that solute rejection, which is given by the ratio of flows, is independent of membrane thickness. The membrane thickness is obviously important, however, in that it determines the flow of water per unit membrane area and this strongly affects processing costs. The rejection increases with the net pressure difference, $\Delta p - \Delta \pi$, because the water flow varies linearly with this difference but the solute flow is nearly independent of it. Rejection is also determined by the ratio $A/B \equiv D_1 c_1 v_1 / D_2 KRT$. Therefore, if $\Delta p - \Delta \pi$ is arbitrarily set at 50 atm, it can be shown that in order to achieve 99% solute rejection, for example, A/B must be 2.0 g/(cm³)(atm) or the ratio of water to solute permeability, $D_1 c_1 / D_2 K$, must be 2700 g/cm³.

B. Irreversible Thermodynamics

Relationships 2 and 3 presented above are essentially specific forms of Fick's law, for which Equation 1 can be considered a general statement. This law states that the flow of each component is proportional to the component concentration gradient, and its validity has been demonstrated in a vast number of experiments. However, in experiments with membranes, particularly biological membranes, the flow of each component is in many cases additionally determined by the flows of the other components; that is, the flows are coupled. This type of behavior is just what is predicted by the thermodynamics of irreversible processes.

Starting with irreversible thermodynamics, Kedem and Katchalsky[10,11] derived a set of relationships describing flow in membranes which are comparable to but more general than Equations 2 and 3 above. For a dilute two-component system, consisting of water and solute, the expressions are

$$J_v = L_p(\Delta p - \sigma\, \Delta\pi) \qquad (10)$$

and

$$J_2 = \bar{c}_{2s}(1 - \sigma)J_v + P\, \Delta c_{2s} \qquad (11)$$

where J_v = the volume flux, which is usually nearly identical to the water flux

L_p = the mechanical filtration capacity of the membrane

σ = the Staverman reflection coefficient, which is a measure of water–solute flow coupling within the membrane

$\bar{c}_{2s}$ = a mean solute concentration in solution*

P = the coefficient of solute permeability at zero volume flow, $P = (J_2/\Delta c_{2s})_{J_v=0}$

The two types of permeability coefficients, L_p and P, contain the membrane thickness, reflecting the fact that these relationships were originally derived for biological membranes whose effective thickness is usually not known.

There is an obvious similarity between Equations 10 and 5 and between Equations 6 and 11. Unfortunately, in the solution-diffusion model the two flows considered are water and solute, whereas in the Kedem-Katchalsky treatment the volume flow and solute flows are considered. This difference is usually numerically trivial but does lead to difficulties in comparing the two sets of permeability coefficients, as has been discussed by Merten.[12] If we ignore these difficulties for the sake of simplicity, when $\sigma = 1$, L_p is nearly identical to A and P is essentially identical to B.† The case of $\sigma = 1$ is for no coupling of water and solute flows within the membrane. When the flows are coupled, $\sigma < 1$, and Equation 10 indicates that the effective osmotic pressure difference is less than the apparent osmotic-pressure difference based on the concentration difference. One possible physical interpretation is that the water drags solute with it as it permeates the membrane. This would have the effect of reducing the water flux in a normal osmosis experiment, for example, and would be equivalent to reducing $\Delta\pi$. Equation 11 indicates that volume flow can cause a flow of solute because of the same water–solute interaction.

In place of the two permeabilities of the solution-diffusion model, three coefficients appear in Equations 10 and 11, L_p, P, and σ, and membrane

* The mean solute concentration has been defined by Kedem and Katchalsky as $\bar{c}_{2s} = \Delta\pi/\Delta\mu_{2s}{}^c$, where $\mu_{2s}{}^c$ is the concentration-dependent part of the chemical potential. For an ideal solution $\bar{c}_{2s}$ has the form $(c'_{2s} - c''_{2s})\ln(c'_{2s}/c''_{2s})$, and for sufficiently small concentration differences $\bar{c}_{2s}$ can be approximated by $(c'_{2s} + c''_{2s})/2$.

† In some of the Kedem papers, a different solute permeability coefficient, denoted by ω, is used.[13] It is defined in this way: $\omega = (J_2/\Delta\pi)_{J_v=0}$. Thus, $\omega\, \Delta\pi = P\, \Delta c_{2s}$.

permeation results expressed in these terms appear in the literature. The flux and solute rejection data of reverse osmosis are readily expressed in terms of these coefficients and the test conditions. Interpretations of these coefficients in terms of the frictional interactions between solute and water, between solute and membrane, and between water and membrane have been made.[13,14] A system of equations analogous to 10 and 11 but with different nomenclature have been derived from irreversible thermodynamics by Johnson, Dresner, and Kraus,[2] who summarize graphically the effects on rejection of water flow, solute exclusion from and mobility within the membrane, and flow coupling.

Based on the Kedem-Katchalsky approach, the solute rejection in terms of the three coefficients is given by[15]

$$SR = \sigma \frac{1 - \exp\left[-L_p(\Delta p - \sigma\,\Delta\pi)(1 - \sigma)/P\right]}{1 - \sigma \exp\left[-L_p(\Delta p - \sigma\,\Delta\pi)(1 - \sigma)/P\right]} \tag{12}$$

or, in terms of the volume flow, by

$$SR = \sigma \frac{1 - \exp\left[-J_v(1 - \sigma)/P\right]}{1 - \sigma \exp\left[-J_v(1 - \sigma)/P\right]} \tag{13}$$

In the high-pressure limit, $J_v \rightarrow \infty$ and $SR \rightarrow \sigma$. The physical significance of this is as follows. At very high volumetric flow, the rate of solute permeation of the membrane by diffusion is trivial relative to the rate of water flow, and in the absence of flow coupling rejection approaches 100%. If the flows are coupled, increased volume flow leads to proportionately increased solute flow and rejection asymptotically approaches a limiting value given by σ. As σ approaches unity, Equation 12 takes the form

$$SR = \frac{L_p(\Delta p - \Delta\pi)}{L_p(\Delta p - \Delta\pi) + P} \tag{14}$$

which is essentially identical to the comparable result of the uncoupled solution-diffusion model (Equation 8).

Regardless of whether the solution-diffusion model or the irreversible thermodynamics system of equations is used, certain general conclusions can be made. First, flow of both water and solute is inversely proportional to thickness. Thus, rejection is not a specific function of membrane thickness. Second, to be effective, the membrane must be highly permeable to the solvent and highly impermeable to the solutes of interest. These permeabilities are determined both by equilibrium characteristics, that is, solubility of water and solute, and by the mobilities within the membrane. In the case of cellulose acetate membranes it has been found that large differences in both solubility and mobility exist between water and most solutes as well as between chemically dissimilar solutes. These effects and their implications have been discussed in detail elsewhere.[2,7,16-18]

C. Pore Model

Ultrafiltration membranes are finely porous structures, and separations occur during ultrafiltration either because solutes are too large to enter the pores or because of frictional interactions within the pores. In the simplest picture of such a membrane, the pores are right-circular cylinders of uniform radius extending completely through the membrane perpendicular to the surface. The water flux through such a membrane is given in terms of the pore radius by Poiseuille's law, one form of which is

$$J_1 = \frac{N \pi r^4 \Delta p}{8 \eta \Delta x} = \frac{\varepsilon r^2 \Delta p}{8 \eta \Delta x} \tag{15}$$

where J_1 = the flux per unit area of membrane
N = the number of pores per unit area
r = the pore radius
η = the viscosity of water
ε = the porosity (ε is equal to $N \pi r^2$ and is usually equated to the water content of the membrane)

This relationship is too simple to adequately describe real membranes for several reasons. Pore tortuosity, blind pores, and the dispersion in pore radii have all been neglected. It is possible to improve the plausibility of Equation 15 by including a tortuosity factor, which could be calculated geometrically from a model of close-packed spheres, for example, and by also including a steric factor to reflect the fact that an approaching molecule can enter a pore only if it does not strike the edges of the pore.[19,20] However, even when these corrections are made and account is taken of heteroporosity this viscous-flow model does not adequately describe most ultrafiltration results. The model predicts that solute retention will not depend on applied pressure or flow rate and, except in those cases where the solute dimension is grossly different from the pore dimension, this is contrary to what is observed experimentally. Further, the pore radii calculated from Equation 15 are not constant when experiments are carried out with different solutes, the calculated radius decreasing with increasing solute size. The calculated radius of pores which retain a significant fraction of the solute is generally very much larger than the molecular radius of the solute as calculated from its density and molecular weight or from the Einstein-Stokes equation.[21]

Real ultrafiltration membranes are not simply sieves that exclude on the basis of size alone, nor are all molecules within the pore fluid carried through the membrane at equal rates by viscous flow without interaction with the membrane. Instead, (a) solutes are excluded in part by those chemical forces that determine solubility (the van der Waal's force in the case of uncharged species and electrostatic repulsion in the case of electrolytes and membranes with fixed charges), and (b) transport within the pore fluid is determined both by viscous flow and by diffusion. The forces that exclude solute in general also affect its transport rate. Thus solute dimensions are only one factor determining ultrafiltration characteristics. These concepts have been recognized for some time. In the early literature, solute exclusion is expressed either in terms of the limited amount of "solvent water" (versus nonsolvent water) within the pore or in terms of the membrane area available for transport of solute. The friction between the molecules and the pore walls was considered theoretically by Faxén.[22,23] The total restriction to ultrafiltration due to steric effects and frictional interaction with the pore walls, is given by the "Ferry-Faxén" equation:[24]

$$\frac{A_a}{A_D} = \left[2\left(1 - \frac{a}{r}\right)^2 - \left(1 - \frac{a}{r}\right)^4\right]\left[1 - 2.104\left(\frac{a}{r}\right) + 2.09\left(\frac{a}{r}\right)^3 - 0.95\left(\frac{a}{r}\right)^5\right] \quad (16)$$

Here, A_a is the pore area available for transport of solute of radius a, A_D is the total cross-sectional area of the pore, and r is the pore radius. The first term in brackets is the steric factor derived by Ferry[20] and the second term is the Faxén friction factor. Pappenheimer[25] and Renkin[26] have incorporated Equation 15 into a transport model in which both viscous flow and diffusion occur within the pores to obtain a consistent set of pore radii for three types of cellophane and a rationalization for retention versus flow rate results.

Merten[7] has developed a "finely porous" model of membrane transport that combines the concept that flow within the pores occurs by both viscous flow and diffusion with the concept of frictional interactions within the pores, first proposed by Spiegler.[14] The water flow through the finely porous membrane is given by

$$u = -\frac{H}{\varepsilon}\left[\frac{dp}{dx} + \frac{f_{23}J_2}{\varepsilon M_2}\right] \quad (17)$$

where u = the velocity of the pore fluid
$\quad H$ = the hydrodynamic permeability ($= \varepsilon r^2 / 8\eta$)
$\quad f_{23}$ = a frictional coefficient equal to the force acting on each mole of solute of molecular weight M_2 as a result of its interaction with the membrane, divided by the solute velocity within the pore (the membrane phase is denoted here by subscript 3)

When this force is small relative to the pressure gradient, Equation 17 reduces to Poiseuille's law, given by Equation 15. The flux of solute per unit membrane area is given by

$$J_2 = \frac{c_{2m}u}{b} - \frac{D_{21}}{b}\frac{dc_{2m}}{dx}$$ (18)

where c_{2m} = the solute concentration expressed as mass per unit total membrane volume

D_{21} = the diffusion coefficient of solute

$$b \equiv 1 + \frac{f_{23}}{f_{21}}$$ (19)

Here f_{21} is a frictional coefficient for solute and water interactions within the pore fluid, and $RT/f_{21} \equiv D_{21}$. Thus $1/b = f_{21}/(f_{21} + f_{23})$ and is equivalent to the Faxén friction factor. When solute–membrane interactions are negligible, $b \cong 1$ and Equation 18 reduces to a mathematical statement of transport via combined bulk flow and diffusion. If solute concentration within the pore is related to external concentration by a distribution coefficient as was done in the solution-diffusion case,

$$c_{2m} = Kc_{2s}$$ (20)

and the concentration reduction factor is given by[7]

$$\frac{c'_{2s}}{c''_{2s}} = \frac{K'' - b\varepsilon + b\varepsilon \exp(u\lambda\,\Delta x/D_{21})}{K' \exp(u\lambda\,\Delta x/D_{21})}$$ (21)

where λ is the tortuosity of the pores. Membrane performance is thus determined by several factors: equilibrium solubility of the solute in the pore fluid, solute–membrane interaction relative to solute–water interaction, and the velocity of the pore fluid, which is given by Equation 17. In the limit of high flow velocities, diffusion is negligible and Equation 21 reduces to

$$\frac{c'_{2s}}{c''_{2s}} = \frac{b\varepsilon}{K'}$$ (22)

When the solute–membrane interaction is small, $b = 1$ and solute retention is determined only by the distribution coefficient for solute in the high flow limit.

As we have noted, irreversible thermodynamics provides a framework for describing membrane transport. This approach is particularly useful when the flows are coupled and one finds the Kedem-Katchalsky system of equations used to describe the behavior of ultrafiltration membranes with increasing frequency. Spiegler and Kedem[15] have carried out an analysis similar to the one presented above based on the Kedem-Katchalsky membrane coefficients L_p, σ, and P. The concentration reduction factor is given (from

Equation 13) by

$$\frac{c'_{2s}}{c''_{2s}} = \frac{1 - \sigma \exp\left[-J_v(1 - \sigma)/P\right]}{1 - \sigma} \tag{23}$$

Although not formally identical, Equations 21 and 23 do have certain similarities. Both have three parameters expressing the three possible inter- actions: water–membrane, water–solute, and solute–membrane. This is in contrast to the two parameters, the water and salt permeabilities, required in the uncoupled solution-diffusion model, where water–solute interactions are assumed negligible. The concentration reduction factor increases with increasing water flow in both cases, approaching $b\varepsilon/K'$ and $1/(1 - \sigma)$, respectively. (The relation between the two types of coupling coefficients, b and σ, has been discussed by Merten.[12]) Furthermore, in both expressions, retention is independent of membrane thickness (in Equation 21, u is propor- tional to $1/\lambda \, \Delta x$) as in the case of the solution-diffusion model.

In general, one cannot differentiate between transport via a solution- diffusion mechanism and transport via a pore-flow mechanism on the basis of the pressure dependence of flow or the solute retention alone. However, flow mechanism should be distinguishable on the basis of temperature dependence. Water flow through porous membranes is characterized by a temperature dependence close to that of the viscosity of water, whereas the temperature dependence of water flow through solution-diffusion type membranes is generally higher.

III. MEMBRANE PERFORMANCE

A. Reverse-Osmosis Membranes

In this section we briefly review the properties of reverse-osmosis mem- branes and cite typical performance characteristics as well as membrane limitations. We begin with the anisotropic cellulose acetate reverse osmosis membrane, which has been explored in considerable detail, and then treat some of the reverse-osmosis membranes developed subsequently.

1. Cellulose Acetate

Most commercially available reverse-osmosis processing equipment utilizes anisotropic cellulose acetate membranes of the general type invented by Loeb and Sourirajan.[5] Some of the properties of cellulose acetate and of the anisotropic membrane have been described.[27] Cellulose acetate combines the three essential requirements for an efficient reverse-osmosis membrane: it is highly permeable to water, its permeability to most water-soluble compounds is relatively quite low, and it is an excellent film-former. Several studies have been made in recent years of the permeability of synthetic polymeric membranes to water and salts in a search for materials with superior properties for water desalination. These studies have confirmed the intuitive expectation that with increasing hydrophilic character, both the water permeability and the salt permeability of membranes increase. A collection of such data is presented in Figure 1 in terms of the permeabilities

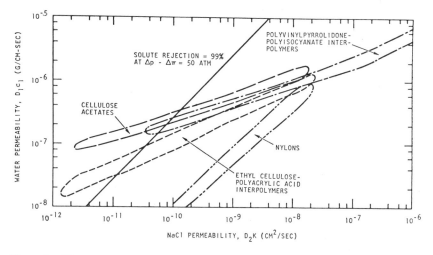

Figure 1. Water permeability vs. NaCl permeability for several types of polymers [28-31]

of the solution-diffusion model. The wide range of values for a given type of material represents changes in chemical composition. For cellulose acetate, for example, the increased water and salt permeabilities result from decreased degree of acetylation. With the polyvinylpyrrolidone–polyisocyanate inter-polymers, the same trend accompanies decreased polyisocyanate content. In addition, the plots are inexact because the permeabilities depend to some extent on the way in which the membranes are cast and on the conditions of measurement. Nonetheless, the trend is clear, and the slopes of the log–log plots are not unity, so that the most selective materials, that is, those with the highest ratio of water to salt permeability, are also those of lowest water permeability. The line drawn with unit slope represents the 99% solute rejection condition, calculated according to the solution-diffusion model with a net pressure difference, $\Delta p - \Delta \pi$, of 50 atm. Cellulose acetates occupy a favorable but not unique position in the plot. A few other materials exhibit comparable intrinsic transport characteristics. With these materials, however, it has not been possible to prepare the effectively very thin films necessary for reverse-osmosis applications.

Loeb and Sourirajan developed a technique (described in the Appendix) for making effectively very thin films of cellulose acetate, which are referred to in this book as anisotropic membranes. The technique has been applied with mixed success to other cellulose esters[32,33] and to ethyl cellulose.[32] However, the method is rather limited in scope, and the understanding of conditions necessary for preparing membrane structures of the Loeb-Sourirajan type is as yet rudimentary.

Thickness.   The importance of membrane thickness is readily illustrated. The thinnest pinhole-free film that is commercially available is about $\frac{1}{4}$ mil, or 6 μ, thick. Even though cellulose acetate is highly permeable to water $(D_1 c_1 \cong 2.5 \times 10^{-7}$ g/(cm)(sec) for the 2.5-acetate), a film of this thickness yields a water flux in reverse osmosis of about 1.5×10^{-5} g/(cm^2)(sec), or 0.3 gal/(ft^2)(day),* at a net pressure difference of 50 atm. A typical anisotropic membrane exhibits equal permselectivity and a water flux of at least 10 gal/(ft^2)(day)† under the same conditions. This demonstrates the fact that the effective thickness of the anisotropic membrane is on the order of 0.2 μ thick.[35] The remainder of the membrane is a highly porous supporting structure with relatively little resistance to water flow and no selectivity to low-molecular-weight solutes. The structure of these membranes has been thoroughly studied with the electron microscope.[36,37] The thickness of the salt-rejecting skin is not clearly resolvable with any of the several techniques used to examine it. Electron micrographs such as Figure 2 show that the skin blends into the porous substructure through a region of graded porosity. Further, the effective thickness is known to depend on the manner in which the membrane is cast and on subsequent treatment. Thus, no definitive thickness can be assigned, and although the thickness can be inferred from water flux, driving force, and water permeability data, some uncertainty in this thickness appears to be inevitable.

The porous substructure of the anisotropic membrane contains about 60% by weight water. If the membrane is allowed to dry without special treatment, it shrinks and distorts markedly and the permeability to water is reduced. Apparently, the pores collapse, increasing the resistance to flow. However, effective methods for drying the membrane without decreasing its permeability have been developed,[39] and the dry membrane has proved to be useful in gas separations.[40]

---

* Multiply flux in g/(cm^2)(sec) by 2.12×10^4 to convert to gal/(ft^2)(day).
† By way of comparison, water production rates in modern distillation plants are typically 10 gal/(day)(ft^2 of evaporator surface).[34]

Figure 2. Electron micrograph of the cross section of an anisotropic cellulose acetate membrane, ×3750.[38]

Annealing Effects.   An important characteristic of the anisotropic cellulose acetate membrane is that a certain amount of tailoring is possible through aqueous annealing. In the as-cast state, the membranes are relatively weak physically, rejection of sodium chloride usually does not exceed 50%, and the membrane constant, A of Equation 5, at low pressure is typically 10×10^{-5} g/(cm²)(sec)(atm) or greater. The membrane can be annealed in water, and with increasing annealing temperature salt rejection increases, the membrane constant decreases, and the mechanical strength improves. The range of typical data is shown in Figure 3. Similar data expressed in terms of sodium chloride permeation constant (B of Equation 6) versus membrane constant are presented in Figure 4. These data are intended to be illustrative only; specific membrane performance is dependent on the composition of

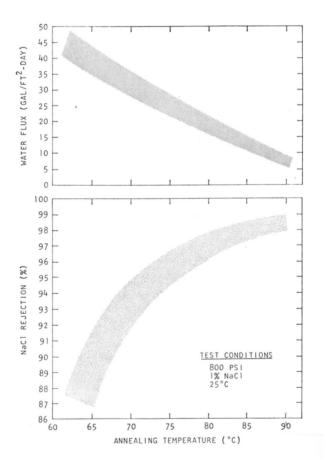

Figure 3.   Typical flux and NaCl rejection of anisotropic cellulose acetate membranes.

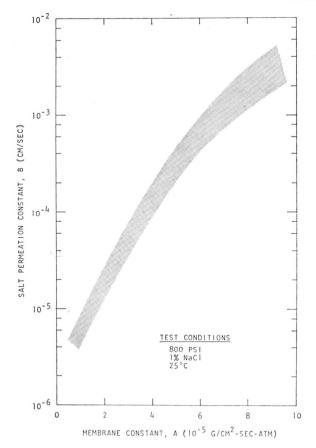

Figure 4. Salt-permeation constant vs. membrane constant for Loeb-Sourirajan type of anisotropic cellulose acetate membranes.[41]

the casting solution, the method of manufacture, and the reverse-osmosis processing conditions. For example, membranes prepared according to the procedure of Manjikian et al.[42,43] have properties similar to those prepared by the original procedure of Loeb and Sourirajan when the former are annealed at about 5°C higher temperature. Some membrane shrinkage occurs on annealing, about 4% in the two planar dimensions at 85°C,[29] and the salt-rejecting skin apparently consolidates and also grows in thickness. Above some temperature in the range 90–95°C, rejection no longer improves even though the membrane constant continues to decrease.

It should be noted that annealing offers only one degree of freedom: rejection cannot be improved without loss in water flux. In fact, rejection, which is determined by the ratio A/B at constant $\Delta p - \Delta \pi$, varies approximately inversely with A over the annealing temperature range 70–90°C, so that the quantity A^2/B is nearly constant at constant operating pressure. This empirical parameter is therefore a good figure of merit for anisotropic cellulose acetate membranes. Typical values fall in the range $1–2 \times 10^{-5}$ when A and B are expressed in cgs units.

The effect of pressure on A and B for membranes annealed at different temperatures is illustrated in Figure 5. The membrane constant is highly pressure dependent in the unannealed state, and the pressure dependence decreases with increasing annealing temperature. The salt-permeation constant is even more strongly dependent on annealing temperature.

In the coupled flow model, the coupling coefficient deviates from unity increasingly with decreasing annealing temperature. Rejection versus flux data reported by Jagur-Grodzinski and Kedem[44] for membranes annealed at different temperatures is presented in Figure 6. The values σ and P are determined for each membrane from flux and rejection data and Equation 13, and are nearly independent of pressure and concentration. The results in both Figures 5 and 6 illustrate the fact that although flow coupling becomes increasingly important with lower annealing temperature, the major effect is a gross increase in salt permeability.

These considerations are generally not encountered in practice, however, because reverse-osmosis processing is usually carried out at a constant pressure, and systems are generally operated at pressures at which the membrane constant is nearly pressure independent; that is, membranes annealed at low temperatures are operated at low pressures. From a practical standpoint the annealing step is a tailoring operation in which membrane properties are fitted to the application. The membrane is heated to the lowest temperature consistent with separation requirements. In this way, the flux is maximized for a given pressure or the pressure is minimized for a given flux, thus reducing either capital or energy costs.

Compaction. At constant operating pressure, the water flux through anisotropic membranes decreases with time. This phenomenon, referred to as membrane compaction, is apparently the result of a creep process in which the thin skin grows in thickness by amalgamation with the finely porous substrate immediately beneath it. It has been found empirically that log (flux) is proportional to log (time) over several orders of magnitude in time. Typical experimental results of membrane constant versus time are shown in Figure 7. The slope of these log–log plots is a function of applied pressure (as shown) and is also a function of membrane heat-treatment temperature as well as the operating temperature. At constant pressure, compaction effects are more pronounced with higher flux membranes. An

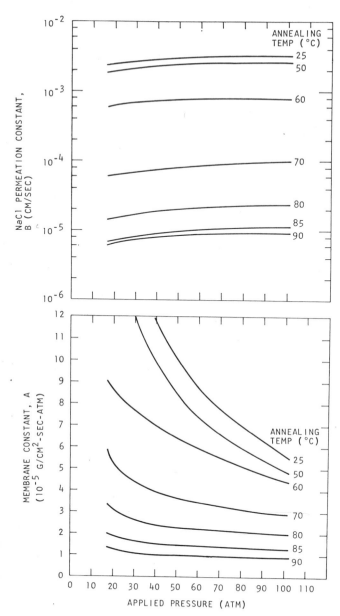

Figure 5. Pressure dependence of salt permeation constant for NaCl and of membrane constant.[41] Loeb-Sourirajan membranes tested at 25°C.

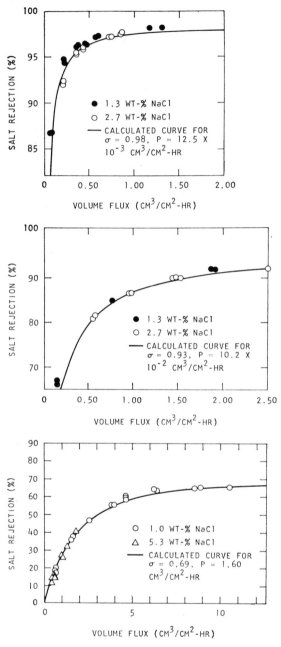

Figure 6. NaCl rejection vs. volume flux for anisotropic cellulose acetate membranes (membrane annealing temperature decreases in the order A, B, C).[44]

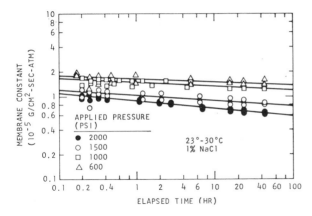

Figure 7. Change in membrane constant with time for anisotropic membranes annealed at 80°C.[29]

extensive series of reliable data is not available, but some values estimated for Loeb-Sourirajan membranes at 25°C and several pressures are summarized in Table 1. The water flux, J_t, at any time t can be calculated from the compaction slope, m, the initial flux, J_0, and the expression

$$J_t = J_0 t^m \tag{24}$$

The calculated flux at one year relative to that at one day is also indicated in the table. Similarly, the average water flux over a given period beginning at one day can be evaluated by integrating this expression with respect to time. The result is

$$\text{average flux} = \frac{J_d \tau^m}{(1 + m)} \tag{25}$$

where J_d is the water flux at one day and τ is the time in days; this expression is valid only for $\tau \gg 1$.

Table 1. Typical Compaction Slopes and Effect on Water Flux for Loeb-Sourirajan Membranes Annealed at 85°C[29,41,45]

Applied Pressure (psi)	Typical Compaction Slope	Flux at One Year/Flux at One Day
500	−0.03	0.84
1000	−0.06	0.70
1500	−0.09	0.59

Several points should be noted before applying these results. First, the slopes are only typical, as noted, and there are substantial membrane-to-membrane variations. Second, significant progress is currently being reported in reducing membrane compaction by modifications to the membrane preparation procedure and by the use of fillers or polymer crosslinking. Third, the observed rate of decline of water flux is frequently greater than indicated by these predictions because of membrane fouling, which is discussed more fully below.

In the absence of other phenomena, solute rejection is generally independent of time: water flux and solute flux both decrease by about the same fraction during compaction. The compaction slope is strongly temperature dependent. At about room temperature, the slope at 1500 psi increases by a factor of two for each 10–15°C increase in temperature.[45]

Fouling. Because of membrane fouling, water-flux-decline rates in field tests have frequently been found to exceed the rate due to compaction. Fouling appears to be common to many membrane processes including, for example, electrodialysis. Details of the mechanism are lacking, but all types of reverse-osmosis membranes and all the engineering configurations are subject to fouling. High feed velocities and low membrane flux reduce but do not eliminate the problem. Fouling has been found to arise from the precipitation of iron and manganese oxides or calcium carbonate and calcium sulfate, and from attachment of organic macromolecules. The latter is particularly prominent in the processing of food products by reverse osmosis (as is discussed in a subsequent chapter) and has been observed in the processing of municipal sewage following conventional primary or secondary treatment. A number of satisfactory solutions have been found in specific cases:[46-49] filtering the feed; maintaining iron and manganese oxides in the reduced forms by excluding air or adding reducing agents; adding precipitation inhibitors such as sodium hexametaphosphate; controlling pH to prevent inorganic precipitates; and passing the feed through an activated carbon column to remove organics. When feed pretreatment has not been practiced or has been ineffective, several techniques can be used for cleaning fouled membranes, including depressurization and back-flushing, and the addition of detergents or acids.

Life. An important characteristic of cellulose acetate membranes is the dependence of water and solute permeabilities on degree of substitution. With increasing degree of acetylation, the solubility of water in the membrane decreases as does the water permeability. The permeability to simple salts and to most other solutes decreases much more rapidly, however, with the result that selectivity increases with increasing degree of substitution. These

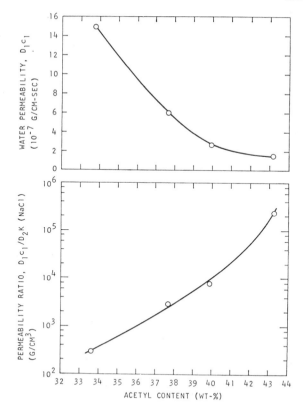

Figure 8.   Permeabilities of cellulose acetate to water and sodium chloride vs. acetyl content at 25°C.[35]

permeabilities have been measured on dense cellulose acetate membranes, and the trends for sodium chloride are indicated in Figure 8 in the D_1c_1, D_2K system. This change in selectivity has an important bearing on membrane life. Cellulose acetate is an ester and is known to hydrolyze in water, and the hydrolysis kinetics are known as a function of pH and temperature.[50] By combining permeability versus acetyl trends with the hydrolysis kinetics data, it is possible to predict the change in water and salt permeabilities and hence salt rejection as a function of time at a given pH and temperature.[51] At either high or low pH, hydrolysis is rapid and the effective membrane life is short. Long-term membrane storage experiments[52] have confirmed, at least qualitatively, the effect of pH on membrane life. This has been demonstrated experimentally in a reverse-osmosis test at high pH, and typical results are

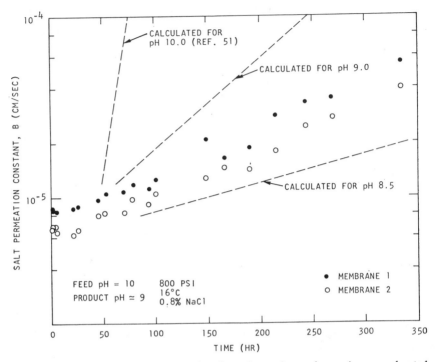

Figure 9. Salt permeation constant vs. time for anisotropic membranes in an accelerated life test.

given in Figure 9[45] in terms of the salt permeation constant, B, for NaCl versus time. The logarithm of the salt-permeation constant is expected to increase linearly with time, and the observations are in accord with this. A pH gradient usually exists across the membrane (because of differences in the rejection of CO_2, HCO_3^-, and CO_3^{2-}), and it appears that membrane life is governed by the pH of the product rather than that of the feed water. Thus the feed water in the experiment depicted in Figure 9 was maintained at about pH 10 with the $NaHCO_3$–Na_2CO_3 buffer system. Because CO_3^{2-} is more highly rejected than HCO_3^-, the product pH was considerably less than 10, the average during the test being about 9. Clearly, membrane degradation is more consistent with the latter pH, as shown by the calculated slopes for three pH values.

The hydrolysis rate is a minimum at pH 4.5–5.0, and membrane life is substantially increased in practice by operating close to this range. If effective membrane life is defined as the time for B to double, the predicted life with NaCl feed is, for example, about 4 yr at pH 4–5 and 2.5 yr at pH 6, but only a few days at pH 1 or 9. There are few experimental data available with which to test the maximum lifetime predictions. In a pilot plant built by the University of California at Los Angeles and operated continuously on brackish water at Coalinga, California, since 1965, membrane life has been excellent even without pH control.[53] After 20 months of continuous operation with a feed and product pH of about 7.5, rejection in one set of membranes fell from 96 to 90%, and some membranes were in use for 3 yr without replacement.[54] The endurance of these membranes considerably exceeded the predictions (which are based, however, on the permeability of dense membranes to sodium chloride). This suggests that the predictions are useful only with highly selective membranes and with sodium chloride feed solutions, and even with these limitations they should not be considered exact. With other membranes such as those used in the Coalinga test and with other feed waters, membrane life is still somewhat unpredictable, although the Coalinga results are highly favorable.

Membrane life can be shortened not only by chemical attack but also by solvation of the membrane skin. For example, if processing of alcohols, organic acids, ketones, aldehydes, or amides is being considered, it should be kept in mind that many of the lower homologs of these polar materials dissolve or plasticize cellulose acetate. Concentrated solutions of certain electrolytes such as zinc chloride and a number of perchlorates also solvate cellulose acetate.

It has been found that at least one strain of microorganism can cause a deterioration in membrane performance, apparently as a result of enzymatic digestion of the cellulose acetate.[55] This strain was cultured in a medium in which cellulose acetate was the essential nutrient. The effect is certainly not general: membranes have been exposed to a variety of microorganisms without apparent attack or deterioration in performance.[52] Chlorination is an effective procedure for disinfecting membranes and reverse-osmosis systems, but excessive exposure to chlorine (e.g., 50 ppm continuously for 10 days) can cause a significant degradation in strength and salt rejection. Dilute cupric sulfate solution has proved to be an effective storage medium: the pH is in the proper range (5.0–5.5) and biological growths are eliminated.

Rejection. The rejection of various solutes by anisotropic cellulose acetate membranes has been examined by a number of groups, both in laboratory experiments and in field tests, and a large body of rejection data exists. The conditions under which the data have been taken vary as do the membrane properties, but the important trends are well established and the data are sufficiently reliable in most cases to determine the feasibility of a specific application. The most extensive single body of data has been collected by Blunk,[56] and a summary has been presented by Loeb.[57] From these and other sources,[58–64] the following generalizations can be made:

1. Rejection increases with increasing ionic charge. Most salts containing divalent ions are virtually completely rejected by membranes annealed at 80°C or higher.

2. In the alkali halide series, rejection decreases as one progresses down the periodic chart; with mineral acids, the trend is reversed.

3. Nitrate, perchlorate, cyanide, and thiocyanate salts are not rejected as well as chloride salts; ammonium salts are not rejected as well as sodium salts.

4. Many low-molecular-weight nonelectrolytes are not well rejected, including solutions of certain gases (ammonia, chlorine, carbon dioxide, and hydrogen sulfide), weak acids such as boric acid, and organic molecules. Among the organics, rejection usually decreases in the order aldehydes > alcohols ≥ amines > acids, and for isomers the order is tertiary > iso > secondary > primary compounds. Within a homologous series, rejection increases with molecular weight, and polyfunctionality tends to enhance rejection. Sodium salts of organic acids are well rejected.

5. Most species with molecular weight greater than about 150, whether electrolytes or nonelectrolytes, are well rejected.

6. Certain substituted phenyl compounds such as phenol and phenol derivatives are negatively rejected (i.e., enriched).

The original literature should be consulted for specific results.

The rejection observed in practice depends on a number of factors. Rejection increases with net pressure, as has already been noted, and the trend of increasing NaCl rejection with increased annealing temperature holds qualitatively for other solutes. The pH of the feed solution can also be important. For example, "combined ammonia" (i.e., $NH_3 + NH_4^+$) rejection is low at high pH, where the equilibrium is shifted toward the neutral species, and the rejection is relatively high at low pH, where the ratio of NH_4^+ concentration to NH_3 concentration is high. Further, the exact composition of the feed solution can be important. When the feed solution is relatively high in divalent ions, the rejection of univalent ions of like sign will tend to be reduced. It has been observed, for example, that when the molar ratio of SO_4^{2-} to Cl^- ions is increased from 0 to 10 with Na^+ being the common cation, the rejection of Cl^- falls from about 98.5 to about 96% at constant $\Delta p - \Delta \pi$.[64]

These factors emphasize the usefulness of membrane tailoring. If the feed solution is high in divalent ions, a membrane annealed at a relatively low temperature could be used in order to achieve acceptable rejection at a lower operating pressure. For example, very-high-flux anisotropic cellulose

$$\delta = \frac{35.3(D_{2s})^{1/3} v^{0.42} d^{1/4}}{\bar{u}_x^{3/4}} \tag{29}$$

acetate membranes can be used in desalting of food products by ultra filtration. These membranes could be made relatively nonselective with respect to salts but still retain sugars, proteins, and other important ingredients

In assessing rejection data to predict the performance of a reverse-osmosis unit, the effect of water recovery should be kept in mind. Most of the rejection data in the literature were obtained under conditions where only a trivial fraction of the water was recovered. In practice, it is generally desired to recover as much water as possible when the application is water desalination (or to concentrate the feed when the application is food processing, for example). One of three effects limits water recovery: the osmotic pressure can reach such high values that it becomes impractical to further concentrate because of membrane compaction, equipment limitations, or energy cost considerations; membrane fouling can become severe because of precipitation of salts or adsorption of colloids from solution; or membrane selectivity can become inadequate. Clearly, for the process to be effective, the fractional rejection must exceed the fractional recovery by a considerable margin. In comparing different reverse-osmosis units or in comparing reverse-osmosis processing with other methods of separation or purification, rejection based on appropriate average concentrations of the feed and concentrate should be used. In most applications, the economies that accompany high recovery will favor high rejection.

It has been reported several times that feed additives are effective in improving salt rejection,[60, 65–67] although their use is not yet widely accepted. The most thoroughly studied of these additives is poly(vinylmethyl ether). The mechanism by which these additives work is not yet clear, although a simple hole-plugging mechanism has been disproved. A number of factors appear to be important in determining the efficiency of feed additives, including feed flow velocity, temperature, the properties of the feed solution and the nature of the additive. The improvement in rejection is accompanied by a reduction in water flux, but the membrane quality parameter, A^2/B, is increased.

Very few solutes have an important effect on either the physical or transport properties of the membrane. The membrane constant is generally not significantly affected by the presence of solutes. Exceptions are the phenols and certain other low-molecular-weight organic species, many of which swell cellulose acetate in dilute aqueous solution; in the presence of these species water flow is generally reduced, in some cases substantially.

A summary of some rejection results[58] is presented in Table 2. The higher rejection and lower water flux associated with higher membrane annealing temperature are apparent. The results with dichlorophenol are noteworthy. Not only was the rejection negative, but the membrane constant was reduced

Table 2.   Summary of Reverse-Osmosis Results[58]

Composition of Solution	Membrane Constant $(10^{-5}$ g/(cm^2)(sec)(atm))	Water Flux (gal/(ft^2)(day))	Rejection (%)
Anisotropic Membranes Annealed at 85°C; Applied Pressure, 1500 psi			
1% NaCl	0.96 ± 0.03	19.1 ± 0.6	98.39 ± 0.07
0.95% NaHCO$_3$	0.90 ± 0.03	18.5 ± 0.5	99.27 ± 0.04
1% NH$_4$Cl	0.92 ± 0.05	18.3 ± 1.0	97.1 ± 0.1
1% NaNO$_3$	0.98 ± 0.02	20.1 ± 0.4	96.4 ± 0.0
98 ppm H$_3$BO$_3$	0.94 ± 0.04	20.2 ± 0.9	56 ± 5
138 ppm urea	0.97 ± 0.02	20.9 ± 0.4	45 ± 1
940 ppm dextrose	1.15 ± 0.06	24.6 ± 1.2	99.74 ± 0.09
0.98% NaH$_2$PO$_4$	1.13 ± 0.06	23.4 ± 1.2	99.93 ± 0.03
106 ppm NH$_4$OH	1.07 ± 0.05	23.0 ± 1.1	25 ± 4
1.5% Na$_2$HPO$_4$	0.95 ± 0.05	19.1 ± 0.9	99.984 ± 0.003
33 ppm 2,4-dichlorophenoxy-acetic acid (2,4-D)	0.87 ± 0.02	18.7 ± 0.4	92.8 ± 0.0
1% NaCl	0.85 ± 0.02	17.2 ± 0.3	98.21 ± 0.21
96 ppm 2,4-dichlorophenol	0.59 ± 0.01	12.7 ± 0.2	−34 ± 2
1% NaCl	0.65 ± 0.01	13.1 ± 0.2	98.08 ± 0.17
Anisotropic Membranes Annealed at 70°C; Applied Pressure, 800 psi			
1% NaCl	2.59 ± 0.13	25.7 ± 1.3	95.1 ± 0.1
1% NaHCO$_3$	2.65 ± 0.15	27.6 ± 1.5	97.9 ± 0.1
1% NH$_4$Cl	2.68 ± 0.13	26.5 ± 1.3	91.0 ± 0.2
1.15% NaH$_2$PO$_4$	2.68 ± 0.13	28.0 ± 1.9	99.82 ± 0.07
155 ppm urea	2.96 ± 0.17	34.0 ± 1.9	27 ± 2
0.85% NaNO$_3$	2.77 ± 0.15	29.5 ± 1.5	88.5 ± 0.8
100 ppm H$_3$BO$_3$	2.99 ± 0.11	34.4 ± 1.3	34.5 ± 0.7
1120 ppm dextrose	3.15 ± 0.04	36.1 ± 0.5	99.38 ± 0.25
1.18% Na$_2$HPO$_4$	2.78 ± 0.06	29.0 ± 0.6	99.94 ± 0.0
90 ppm NH$_4$OH	2.85 ± 0.06	32.8 ± 0.7	7.4 ± 0.7
1% NaCl	2.75 ± 0.05	27.1 ± 0.6	95.3 ± 0.2

irreversibly as evidenced by the tests performed with NaCl before and after the dichlorophenol test.

The importance of water–solute flow coupling has been examined in only a very limited number of cases. It has been shown to be important with highly annealed membranes in the case of phenol solutions,[59] and as noted above, it becomes increasingly important with decreased annealing temperature.

Operating Temperature Effects.   Water flux increases with operating temperature with an activation energy of approximately 5–6 kcal/mole; that is, at about 25°C flux increases by about 3%/°C. The temperature dependence appears to be somewhat dependent on the membrane casting formulation, the membrane annealing temperature, and the nature and concentration of the solutes present;[35, 68, 69] with these parameters fixed, the activation energy is independent of pressure.[69] The activation energy for salt transport is dependent on the nature of the salt, but it generally exceeds that for water transport so that solute rejection improves with decreasing operating temperature. Some data for 0.5-M solutions at 1500 psi are presented in Table 3.[69] To illustrate these data, for NaCl the flux increased from 4.0 to 12.5

Table 3. Activation Energy for Water and Salt Transport Through Anisotropic Cellulose Acetate Membranes[69]

	NaF	NaCl	NaBr	NaI	KF	KCl	KI
Water flux at 25°C (gal/(ft²)(day))	8.1	9.3	8.6	8.5	10.5	8.4	8.8
Salt rejection at 25°C (%)	99.1	97.0	94.9	89.2	98.9	95.4	86.6
Activation energy for water transport (kcal/mole)	5.1	5.4	5.0	5.1	4.8	4.8	4.9
Activation energy for salt transport (kcal/mole)	8.2	7.4	6.4	6.1	6.6	5.6	5.3

gal/(ft²)(day) and the rejection fell from 97.8 to 96.8% as the operating temperature was raised from 0 to 35°C.

2. Other Cellulose Esters

As a result of the program of the Office of Saline Water to develop reverse-osmosis membranes suitable for seawater desalination, several membranes have been developed with improved selectivity with respect to salts. All these are cellulose ester membranes: a blend of cellulose diacetate and cellulose triacetate;[70, 71] a highly substituted cellulose acetate;[70, 71] cellulose acetate methacrylate;[70, 71] cellulose acetate butyrate;[72] and a cellulose acetate "thin-film composite" membrane.[64, 73, 74] All these membranes have a degree of substitution (DS) of 2.6–2.8, which may be compared with a DS somewhat lower than 2.5 for the 39.8% acetyl cellulose acetate from which the membranes described in the previous section were prepared. The difference in DS may appear rather small, but, as has been noted, the transport properties are strongly dependent on DS as are the solubility properties.

All these membranes except the last one are anisotropic and are prepared by essentially the Loeb-Sourirajan technique, suitably modified to compensate for the different solubility characteristics. The thin-film composite membrane is prepared by applying a very thin film of cellulose acetate, typically 500–1000 Å thick, to a finely porous support membrane. The salt rejection performance characteristics of all these membranes are similar: rejections of NaCl are frequently in excess of 99.5 %. Water fluxes are generally lower than those attainable with lower selectivity cellulose acetate membranes: membrane constants are in the range $0.4–0.8 \times 10^{-5}$ g/(cm²)(sec)(atm).

The cellulose acetate blend membrane has been tested with seawater feed for several months with favorable results. The product water was potable, based on the U.S. Public Health Service potability standard of 500 ppm total dissolved solids, over a 6-week period, and the water flux was in the range 6–8 gal/(ft²)(day) at an operating pressure of 1500 psi. Salt rejection was initially 99.4% and the water recovery was somewhat less than 15%.[49]

Some results for thin-film composite membranes are presented in Table 4.[64]

Table 4. Reverse-Osmosis Results for Thin-Film Composite Membranes[64] (Test Conditions: 1500 psi, 25°C, 1% NaCl)

	650-Å-Thick Cellulose 2.83-Acetate Thin Film	950-Å-Thick Cellulose 2.65-Acetate Thin Film
Water flux at 1 hr (gal/(ft²)(day))	9.3	10.7
Final water flux (gal/(ft²)(day))	7.3	9.2
Final membrane constant (10⁻⁵ g/(cm²)(sec)(atm))	0.37	0.46
Final NaCl rejection (%)	99.86	99.59
Time of test (hr)	190	124
Compaction slope	−0.046	−0.031

3. Hollow Fibers

Hollow-fiber membranes represent a distinct approach to reverse-osmosis separation devices. As we have noted, the essential feature of the anisotropic cellulose acetate membrane is the presence of a very thin solute-rejecting skin. The high membrane fluxes that result allow the design of devices in which high production rates per unit of pressure vessel volume are achievable. The designs are along conventional lines in which membranes are held in a plate-and-frame configuration, in a tubular configuration, or in a modified plate-and-frame in which the membrane is rolled into a spiral-wound module. With membrane fluxes of 10–20 gal/(ft²)(day) and membrane packing densities as high as 200 ft²/ft³ of pressure-vessel volume, specific production rates of 2000–4000 gal/(day)(ft³ of pressure-vessel volume) are achievable in water desalination. In the hollow-fiber approach, the membranes are isotropic and nonporous, and high membrane flux is sacrificed for very high membrane packing density to achieve specific water-production rates of the same order of magnitude. Typical fiber dimensions are: 45-μ OD and 24-μ ID. The ratio of outside diameter to wall thickness is sufficiently small that the fibers are capable of withstanding high external pressure without additional support. However, even in the best case the wall thickness is more than a factor of ten greater than the effective thickness of the anisotropic cellulose acetate membrane and fluxes are reduced accordingly. On the other hand, membrane packing densities on the order of 10,000 ft²/ft³ of pressure-vessel volume are typical (for example, 28 million fibers can be packed into a 14-in. Schedule 40 pipe),[75] so that with typical fluxes of 0.1 gal/(ft²)(day), water-production rates of 1000 gal/(day)(ft³ of pressure-vessel volume) are achievable.

Unfortunately, very few reports of the preparation and performance of hollow-fiber membranes have appeared in the open literature, so sufficient information is not available for an overall performance comparison between hollow-fiber systems and those based on the anisotropic cellulose acetate membrane.

Hollow fibers have been successfully prepared from nylon[76] and cellulose acetate.[77-79] In both cases, the observed solute rejections are considerably less than those commonly achieved with anisotropic cellulose acetate membranes. Water flux is an inverse function of rejection, at least with the nylon hollow fibers, and membrane tailoring has been possible, as is illustrated in Figures 10 and 11.[76] A comparison of specific ion rejections attained in field tests with Loeb-Sourirajan type membranes,[47] nylon hollow fibers,[76] and cellulose acetate hollow fibers[79] is presented in Table 5. A direct comparison of this type can be misleading, particularly when the feed solutions are of different

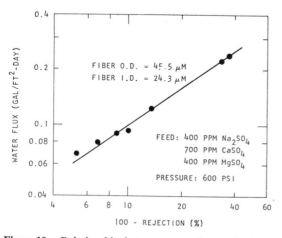

Figure 10.   Relationship between water flux and rejection of mixed sulfates for nylon hollow fibers.[76]

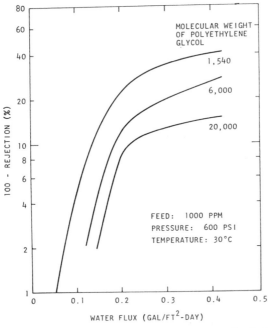

Figure 11. Relationship between water flux and rejection of polyethylene glycols for nylon hollow fibers.[76]

composition. For the anisotropic membrane, the feed was largely NaCl, whereas in the other two cases the feed was largely sulfate salts. As Table 5 indicates, the sulfate ions are highly rejected by all these membranes, and because of the requirement of charge neutrality an equal number of equivalents of counterions, that is, cations, are rejected with them. One would therefore expect that had all three membranes been tested on the same feed water, the anisotropic cellulose acetate membrane would have compared even more favorably in the rejection of total dissolved solids.

The order of rejection for the nylon hollow fiber appears to be similar to that previously given for cellulose acetate; for example, divalent ions are rejected better than monovalent ions. The selectivity of nylon hollow fibers is apparently pH sensitive. In tests at Firebaugh, California, the rejection of most of the ions improved when the pH of the feed brine was lowered from 7.4 to 5.8. These higher rejections are reported in Table 5.

Fouling problems have been encountered with hollow-fiber membranes similar to those observed with anisotropic cellulose acetate membranes; that is, suspended matter becomes trapped in the fiber bundle, and colloidal organic and inorganic matter as well as sparingly soluble salts become

Table 5. Rejection of Specific Ions with Three Reverse-Osmosis Systems

	Anisotropic Cellulose Acetate Membrane[a] [47]	Nylon Hollow Fiber[b] [76]	Cellulose Acetate Hollow Fiber[79]
Total dissolved solids in feed (ppm)	4580	6000	No data
Temperature (°C)	20	24	No data
Pressure (psi)	600	600	No data
Feed pH	6.0	5.8	No data
Water recovery (%)	75	24	50
Rejection[c]			
Na^+	95.1	89	83
Cl^-	96.6	57	23
SO_4^{2-}	100	97.4	98.7
Ca^{2+}	99.1	97.9	96.4
Mg^{2+}	99.9	98.3	89
SiO_2	90	35	73
Boron	53	0	No data
NO_3^-	87	21	No data
HCO_3^-	98.4	51	No data
K^+	93.4	87	No data
F^-	88	>50	No data
Iron	99.2	70	No data
Total hardness (as $CaCO_3$)	99.6	97.7	98.4
Total alkalinity (as $CaCO_3$)	98.4	51	96.9
Total dissolved solids	97.2	90.5	No data

[a] River Valley Golf Course test of a 50,000-gal/day plant, San Diego, California.
[b] Test at Firebaugh, California, with adjustment of pH to 5.8.
[c] Based on a simple average of feed and reject brine concentrations.

attached to the membrane and reduce water flux. High feed velocities and feed pretreatment appear to minimize these effects.

Because of the high membrane packing density available, the hollow-fiber approach may offer certain advantages in processes other than reverse osmosis. For example, an artificial kidney machine using regenerated cellulose hollow-fiber membranes has been developed and has been evaluated clinically.[80] Also, since the choice of membrane materials is not restricted to those capable of forming an anisotropic membrane, hollow fibers, such as the nylon now used, may be compatible with solutions that are incompatible with cellulose acetate. For example, the rejection of H_3BO_3 by nylon hollow fibers was found to increase with increasing pH above pH 8,[75] apparently because the $H_2BO_3^-$ ion is highly rejected. The rejection was quite high at pH

11; at this pH the useful life of cellulose acetate membranes is very short. (Operation at such high pH might not be practical in some applications because of the possibility of $CaCO_3$ precipitation.)

4. Other Reverse-Osmosis Membranes

McKelvey, Spiegler, and Wyllie[81] showed that conventional ion-exchange membranes were effective in desalination by reverse osmosis. The rejection was a function of the concentration of the feed solution, and it was higher in the case of divalent co-ions (ions of the same sign as the fixed charge on the membrane) than for monovalent co-ions; both findings are in agreement with the Donnan ion-exclusion principle.[82] However, the water fluxes observed by McKelvey and co-workers were uninterestingly low. This is due in part to the fact that the water permeability of ion-exchange membranes is low, and in addition the membranes are effectively very much thicker than anisotropic cellulose acetate membranes.

Only the anisotropic cellulose acetate membranes and hollow fibers are used in commercially available reverse-osmosis systems. A limited number of promising alternative membrane materials have been developed under the Office of Saline Water desalination program. Among these are dynamically formed membranes, graphitic oxide membranes, and porous glass membranes. The properties and current state of development of these membranes have been described in Office of Saline Water annual reports (e.g., Reference 83) and in numerous publications.[84-90]

B. Ultrafiltration Membranes

Synthetic membranes have been in use for over 100 yr for concentration and separation by ultrafiltration. Prior to about 1960, there were very few industrial uses for the process, and most workers carried out small-scale processing either to isolate or concentrate desired products or to study the fundamentals of membrane permeation. The field prior to 1935 has been thoroughly reviewed by Ferry.[20]

The early work was limited for the most part to studies with cellophane or porous cellulose nitrate films, the latter being cast from either an acetic acid solution[91] or an ether–alcohol "collodion" solution.[92] Because the cellulose nitrate films were individually prepared by each investigator, reproducibility was poor and comparisons between different studies were tenuous. Adsorption of solute on the pore walls and membrane plugging were commonly observed. The finest pore sizes produced by these methods were in the range 20–30 Å, and by modification of the technique, membranes with pore radii of 1 μ or greater were prepared. It is interesting to note that rejection of low-molecular-weight salts was observed with certain types of cellophane and cellulose nitrate membranes;[20, 93, 94] the earliest record of desalination by a pressure-driven membrane process appears to be 1926, although it was doubtlessly achieved even earlier.

Depending on the transport model used, the properties and performance of ultrafiltration membrane are variously described in terms of the parameters introduced earlier: effective pore radius, water permeability, solute retention, and a water–solute coupling coefficient such as the Staverman reflection coefficient. A summary of the various methods used to calculate the radius

Table 6. Flow Characteristics of Three Types of Cellophane Membranes[95]

Substance	Molecular Weight	Molecular Radius (Å)	Reflection Coefficient, σ		
			Dialysis Tubing[a]	Cellophane[b]	Wet Gel[c]
D_2O	20	1.9	0.002		0.001
Urea	60	2.7	0.024	0.006	0.004
Glucose	180	4.4	0.20	0.044	0.016
Sucrose	342	5.3	0.37	0.074	0.028
Raffinose	595	6.1	0.44	0.089	0.035
Inulin	991	12	0.76	0.43	0.23
Bovine serum albumin	66,000	37	1.02	1.03	0.73
Membrane constant, A or L_p (10^{-5} g/(cm²)(sec)(atm))			1.7	6.5	25
Calculated pore radius (Å)			23	41	82

[a] Visking cellulose.
[b] Du Pont 450-PT-62 cellophane.
[c] Sylvania 300 viscose wet gel.

of the pores in ultrafiltration membranes is presented by Lakshminarayanaiah;[24] the Poiseuille equation, Equation 15, invariably yields the smallest value. The mean pore radius of cellophanes ranges from about 15–25 Å (depending on the method used to calculate the radius) to 80–100 Å; the radius increases with water content. A summary of Staverman reflection coefficients for several solutes, pore radii, and membrane constants for three types of cellophane film is presented in Table 6.[95] Retention increases with molecular weight of the solute and with decreasing pore size, as anticipated, and there is significant retention of solutes by membranes whose mean pore radius is several times that of the solute molecule. The range of membrane constants of these cellophane films is similar to that commonly observed with anisotropic cellulose acetate membranes that have been subjected to a range of annealing temperatures.

Data of this type have been used to estimate solute–membrane and solute–water interactions in a few cases. The frictional coefficient (f_{21} in Equation 19) for solute–water interactions is usually considerably larger than that for solute–membrane interactions in ultrafiltration membranes, and the ratio of these two coefficients increases with increasing water content. That is, the pore approaches an open pipe and the water–solute interaction approaches its free solution value as the water content increases. These interactions can be measured only when significant solute is present in the pore fluid, that is, when the solute is not highly retained. Some results are presented in Table 7.[96] The solute–water interaction within the membrane appears to be substantially greater than the free-solution value, $f_{21,s}$ (calculated from free-solution diffusion coefficients and the Einstein equation, $D_{21,s} = RT/f_{21,s}$). However, the ratio $f_{21}/f_{21,s}$ can be interpreted as the tortuosity of the membrane pores. Merten[7] has analyzed the data of Henderson and Sliepcevich[97] for ultrafiltration of sucrose solutions through cellophane in terms of the finely porous membrane model; he arrived at a value for the tortuosity similar to that in Table 7 for sucrose, that is, $\lambda = 6.4$.

Beginning around 1960, new types of ultrafiltration membranes became commercially available. These membranes have largely replaced cellophane and porous cellulose nitrate in laboratory and batch-scale work, and several industrial applications involving continuous ultrafiltration processing have been developed. Taken collectively, all the available types of membrane now comprise a family of membranes covering an extremely broad range of permeability and retention characteristics. Although there are some limitations to their use, a near-continuum exists in available membrane characteristics, from reverse-osmosis membranes that are highly impermeable to sodium chloride to ultrafiltration membranes that pass macromolecules with molecular weights of 10^6 or greater. Furthermore, membrane fabrication technology is now sufficiently advanced that the remaining gaps in performance may soon be filled. These new membranes are highly reproducible relative to the collodion ultrafiltration membranes of the 1930's. Some of the new membranes are anisotropic, or "skinned," similar to the Loeb-Sourirajan cellulose acetate membrane, so that the flux per unit pressure difference is much greater than that for collodion or cellophane films of equal selectivity.

A partial list of the commercially available membranes, including reverse-osmosis membranes, is presented in Table 8. In most cases water fluxes in Table 8 have been normalized to 100 psi, but not all membranes can be used at this pressure. The manufacturer's recommended operating pressure for the UM series is below 100 psi and for the XM-50 and XM-100 membranes is below 50 and 25 psi, respectively. Because of concentration polarization effects, membrane fouling and clogging, or membrane instability at high pressure, the most permeable membranes are generally operated at low

Table 7. Reflection and Friction Coefficients for Dialysis Tubing[96]

Solute	Concentration (mole/l)	σ	f_{21} (10^{15} dyne-sec/(mole)(cm))	$f_{21}/f_{21.s}$	f_{23} (10^{15} dyne-sec/(mole)(cm))	f_{13} (10^{15} dyne-sec/(mole)(cm))
H$_2$O	—	0.006	3.3	3.2	—	—
Urea	0.5	0.006	6.6	3.8	0.65	0.083
Glucose	0.0125	0.12	18	4.8	3.1	0.084
Glucose	0.025	0.11	18	4.9	2.9	0.085
Glucose	0.050	0.083	19	5.1	2.3	0.085
Glucose	0.10	0.072	20	5.2	2.1	0.087
Sucrose	0.0125	0.16	31	6.5	7.3	0.084
Sucrose	0.025	0.14	33	6.7	6.5	0.086
Sucrose	0.50	0.12	34	6.9	5.9	0.087
Sucrose	0.10	0.11	36	6.9	5.7	0.090

Table 8. Commercially Available Ultrafiltration and Reverse-Osmosis Membranes[a]

Membrane	Manufacturer	Type	Chemical Composition	Net Pressure (psi)	Water Flux (gal/(ft²)(day))	Solute	Rejection (%)	Principal Use
Loeb-Sourirajan, 85°C annealed	Several	Anisotropic	Cellulose acetate	1000	20	NaCl	97–98	Water desalination
Loeb-Sourirajan, 70°C annealed	Several	Anisotropic	Cellulose acetate	400	20	NaCl	90	Water desalination
Loeb-Sourirajan, unannealed	Several	Anisotropic	Cellulose acetate	150	20	NaCl	25	Water desalination
Gel cellophane	Du Pont, Union Carbide	Homogeneous	Cellulose	100	1.5	Sucrose	15	Concentration and demineralization of proteins and organic solutes
P.E.M.	Gelman	Isotropic	Cellulose triacetate	100	8.5	10,000-MW dextran; 67,000-MW BSA[b]	0; 100	Concentration and demineralization of proteins and organic solutes
P	Schleicher & Schuell	Homogeneous	Cellulosic	100	35	10,000-MW dextran; 67,000-MW BSA	0; 90	
Diaflo UM-3	Amicon	Anisotropic	Polyelectrolyte complex	100	25	Sucrose	90	Concentration and demineralization of proteins and organic solutes
Diaflo UM-2	Amicon	Anisotropic	Polyelectrolyte complex	100	60	Sucrose	50	Concentration and demineralization of proteins and organic solutes
Diaflo UM-1	Amicon	Anisotropic	Polyelectrolyte complex	100	175	6000-MW PEG[c]	80	Protein concentration, purification
Pellicon PSED	Millipore	Anisotropic	Cellulose ester	100	40	20,000-MW trypsin	50	Concentration, purification of aqueous solutions
Pellicon PSAC	Millipore	Anisotropic	Cellulose ester	100	120	Sucrose	40–60	Concentration, purification of aqueous solutions
HFA 100	Abcor		Cellulosic	100	25	10,000-MW dextran	100	Macromolecular concentration, purification
HFA 300	Abcor		Cellulosic	100	600	13,500-MW cytochrome	20	Macromolecular concentration, purification
Diaflo XM-50	Amicon	Anisotropic	Substituted olefin	100	350	67,000-MW BSA	100	Macromolecular concentration, purification
Diaflo XM-100	Amicon		Substituted olefin	100	1,050	160,000-MW gamma globulin	0	Macromolecular concentration and removal
Diaflo PM-10	Amicon		Aromatic polymer	100	1,050	1400-MW bacitracin	100	Macromolecular fractionation
Diaflo PM-30	Amicon		Aromatic polymer	100	2,800	35,000-MW pepsin; 110,000-MW dextran	100; 70	Protein concentration, purification
VF	Millipore	Isotropic	Cellulose ester	100	1,000	250-Å pores	100	Macromolecular concentration, purification
HA	Millipore	Isotropic	Cellulose ester	100	15,000	4500-Å pores	60	
Nuclepore	General Electric	Isotropic	Polycarbonate	14	34,000	0.5 ± 0.06 μ pores		Bacteria removal

[a] Taken largely from Michaels.[1]
[b] BSA = bovine serum albumin.
[c] PEG = polyethylene glycol.

pressure, and typical operating fluxes in many cases are <50 gal/(ft²)(day). Several potential applications of ultrafiltration membranes have been sited.[98]

For most of the membranes in Table 8 other than cellulose acetate, the details of preparation are proprietary. The polyelectrolyte complex membranes are formed[99, 100] by reaction between a polycation, such as poly-(vinylbenzyltrimethylammonium chloride), and a polyanion, such as sodium poly(styrenesulfonate). The anisotropic membranes prepared from these complexes are apparently similar in structure to the anisotropic cellulose acetate membrane: a relatively dense skin 0.1–10 μ thick is supported on a

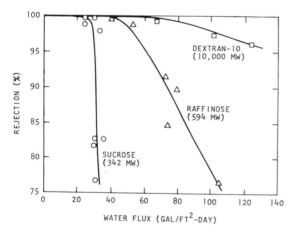

Figure 12. Variation of solute rejection with water flux for a series of anisotropic polyelectrolyte complex membranes.[1]

porous substructure 20–100 μ thick. Membrane tailoring comparable to what is accomplished with anisotropic cellulose acetate is possible with these membranes. Figure 12 is a plot of rejection of three polysaccharides as a function of flow rate.[1] Each data point for a given compound was obtained with a different membrane.

There appear to be no published detailed studies of transport mechanisms in these new ultrafiltration membranes. Michaels has reported[1] that some of the Amicon membranes are microporous, whereas others appear to operate as diffusive barriers in which both diffusion and viscous flow of solute are important. The performance of the two types is quite different. The microporous membranes foul and plug more easily, they are less discriminating with regard to solutes of similar size, and solute retention does not increase (and can, in fact, decrease) with increasing applied pressure. With "diffusive" ultrafiltration membranes, retention increases with pressure.

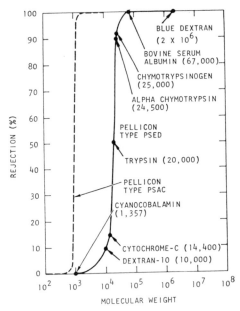

Figure 13. Rejection vs. molecular weight for two types of Pellicon membranes.[101]

The microporous membranes are for the most part isotropic (i.e., they are not "skinned"), and they clog easily when used to ultrafilter solutions with a continuous spectrum of solute size, such as polymer solutions. The Pellicon membrane appears to be an anisotropic microporous membrane that is more permeable to water and more retentive with less tendency for clogging than previous microporous membranes. A plot of the retention characteristics of two Pellicon membranes is presented in Figure 13. The molecular weight "cutoff" for these membranes is sharp relative to most ultrafiltration membranes, a reflection, apparently, of a relatively isoporous structure. The pores in the General Electric Nuclepore membrane filters are introduced by etching radiation damage sites. Porosity is relatively low, 5–15%, but the pores are quite uniform in dimensions and are apparently straight-through cylindrical holes. Water flow through the 0.5-μ membrane is in good agreement with Poiseuille's law.

IV. CONCENTRATION POLARIZATION

As water is removed through the membrane in both reverse osmosis and ultrafiltration, the rejected species tend to accumulate at the solution–membrane interface. These species are transported away by diffusion, so that at steady state a concentration gradient exists at the interface in which the transport of these species toward the membrane by the bulk solution flow is balanced by the combined effects of diffusive flow in the opposite direction and permeation through the membrane. There are three negative effects associated with this concentration polarization:

1. The increase in concentration of solute* at the interface leads to an increase in solute flow through the membrane. When the membrane permeability is not a function of the external solution concentration (as is generally the case with cellulose acetate membranes), solute flow increases linearly with the interfacial concentration.

2. The increase in local concentration can lead to saturation in one of the solution components, which may result in precipitation or gelation of that component on the membrane surface, reducing the effective membrane area or effectively adding a second "membrane" in series. The water flow decreases accordingly.

3. The interfacial osmotic pressure increases, resulting in a decrease in $\Delta p - \Delta \pi$ and a decrease in water flow.

In the desalination of brackish waters, it usually proves to be economically advantageous with existing reverse-osmosis membranes to operate with $\Delta p \gg \Delta \pi$, so that the third effect is of minor consequence in this case and water flux can be maintained with only slight relative increase in Δp. In sea-water desalination or in the processing of certain foods with high osmotic pressures, this is not the case.

Although the polarization effect cannot be completely eliminated, there are two approaches to minimizing it: (*a*) operate at low water flux or with very narrow feed channels in order to reduce the concentration gradient that develops at a given fractional recovery of solvent, and (*b*) make the flow of feed solution turbulent in order to continually mix most of it. The choice between these solutions depends on a number of economic factors, the most important of which is the relative cost of membranes and pumping energy.

The mass transport equation expressing the steady-state condition within the polarization region at any point along the feed flow channel is

$$\frac{J_1 c_{2s}}{c_{1s}} - D_{2s} \frac{dc_{2s}}{dx_s} = B \, \Delta c_{2s} \qquad (26)$$

where $D_{2s} =$ the solute diffusion coefficient in the solution
$x_s =$ the distance parameter in solution perpendicular to the membrane surface

* In the discussion that follows, we use the term "solute" for simplicity, but it should be recognized that polarization effects occur for colloidal and suspended matter as well.

The first term is the rate of solute transport to the membrane surface resulting from bulk flow toward the membrane; the second term represents the diffusive flux back into the bulk of the solution, assumed to obey Fick's law; and the right-hand side expresses the flow of solute through the membrane. When the membrane is solute-impermeable, $B = 0$ and the equation can be integrated to give

$$\frac{c'_{2s}}{c_{2s}{}^{b}} = \exp\left(\frac{J_1 \Delta x_s}{c_{1s} D_{2s}}\right) \tag{27}$$

where $C_{2s}{}^{b}$ = the "bulk" solute concentration far from the membrane surface
c'_{2s} = the interfacial concentration
Δx_s = a characteristic polarization distance in the solution

In the laminar flow regime the polarization distance increases with distance down the feed channel (or with time in the absence of feed flow), until eventually the polarization region extends over the entire feed channel height. When the flow is turbulent, most of the feed channel is well mixed and, according to the film theory model of heat and mass transfer, the polarization region extends only over a boundary layer adjacent to the membrane surface, the thickness of which is determined by the linear velocity, channel geometry, and solution properties.

In processing feeds containing solutes of low diffusivity, the polarization problem can be very severe, regardless of whether feed flow is laminar or turbulent. As shown in Figure 14, the flux through three types of ultra-filtration membranes leveled off with increasing pressure at values well below the fluxes achieved with pure water. Macromolecules and biocolloids present in the plasma being ultrafiltered collected at the membrane surface and

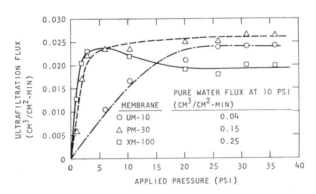

Figure 14. Ultrafiltration of human plasma in a stirred cell.[102] Limiting fluxes are 7–9 gal/(ft²)(day).

formed an adherent gel layer adjacent to the membrane. Similar phenomena have been observed in the concentration of foods with ultrafiltration or reverse-osmosis membranes.

Equation 26 has been solved for a variety of conditions in both the laminar and turbulent flow regimes by Brian and co-workers[103–105] and by Gill and co-workers.[106] In spite of the fact that several simplifying assumptions are generally made, these solutions become quite complex because the problem is basically nonlinear. The water flux through the membrane decreases down the channel (even if the effect of frictional losses on Δp is ignored), because $\Delta \pi$ is constantly increasing as water is removed and as the boundary layer develops. Solute flux increases for the same reasons. As a result of these complexities, the results do not appear in closed form, at least in the laminar flow case, and finite-difference methods have been used. These results are available elsewhere[2, 105] and are not repeated here. We instead cite only the essential features of the solutions, note the conditions where either laminar or turbulent flow seems most appropriate, and give the polarization modulus, $c'_{2s}/c_{2s}{}^b$, in certain interesting cases.

A. Laminar Flow

As noted above, in laminar flow the concentration gradient develops over the entire height of the feed channel. However, depending on flow conditions and the diffusivity of the solute in solution, a large fraction of the water may be removed before the gradient is fully developed. Thus there are two different laminar flow regimes: the entrance region where the concentration gradient increases down the length of the channel, and an asymptotic region where the gradient extends across the entire flow channel and does not change shape as additional water is removed.

In the entrance region the polarization modulus depends on a combination of two dimensionless quantities: a mass transfer coefficient, $J_1 d/D_{2s}$ (the Péclét number), where d is the half-height of a channel between two parallel flat membranes; and the solvent recovery fraction, which is given by the quantity $J_1 y/\bar{u}_x d$, where y is the distance down the channel and $\bar{u}_x$ is the feed velocity averaged over the channel height. Figure 15 shows the degree of polarization, expressed as $\Gamma = (c'_{2s}/c_{2s}{}^b) - 1$, plotted against the parameter

$$ \zeta = \frac{1}{3}\left(\frac{J_1 d}{D_{2s}}\right)^2 \left(\frac{J_1 y}{\bar{u}_x d}\right) $$

for the entrance region, along with three asymptotic values.[2] Complete solute rejection and constant flux through the membrane were assumed in these calculations, but polarization values have been calculated for several

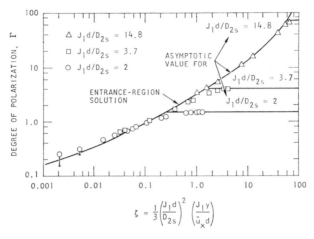

$$\zeta = \frac{1}{3}\left(\frac{J_1 d}{D_{2s}}\right)^2 \left(\frac{J_1 y}{\bar{u}_x d}\right)$$

Figure 15.   Polarization in laminar flow.[2]

cases of incomplete rejection and for a flux dependent on $\Delta p - \Delta\pi$. Polarization decreases with decreasing rejection, but the average polarization over the channel length is nearly the same in the constant-flux case as in the variable-flux case. Solutions to the polarization problem have also been obtained for tubular membranes.[105]

Far downstream in the asymptote region, the polarization is reasonably well represented by $\Gamma = \frac{1}{3}(J_1 d/D_{2s})^2$. For $\Gamma < 1$, this approximation begins to fail, but the exact solution has been given.[104, 105] Figure 16 is a plot of Γ versus $J_1 d/D_{2s}$ showing the entrance region solution for various recoveries and the far-downstream solution to which the entrance-region solutions converge. For a given diffusivity, feed velocity, and flux, the distance from the channel entrance to the point of asymptotic flow can be estimated for various recoveries from this plot. It is interesting to note that for a fixed fractional recovery, the average polarization varies with channel height and flux but not with feed velocity. The feed velocity affects the polarization at each point along the length, but the length required for a given recovery increases with increasing velocity and the average polarization is fixed. It is possible to carry out a process entirely in the entrance region if multiple passes are combined with low recoveries per pass or, equivalently, if devices are used to induce mixing periodically down the length of the channel. This approach has been proposed by Ginnette and Merson for food processing and is described in detail in Chapter 10.

The feasibility of reverse-osmosis or ultrafiltration processing in the laminar flow regime can be estimated from Figure 16. If we arbitrarily set as an acceptable upper limit $\Gamma = 1$ and require reasonable recovery per pass, then

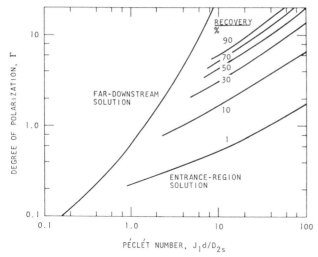

PÉCLÉT NUMBER, $J_1 d/D_{2s}$

Figure 16. Degree of polarization Γ vs. Péclét number for various water recoveries in laminar flow.

most of the recovery occurs in the far-downstream region. For anisotropic cellulose acetate membranes, a typical J_1 is 10 gal/(ft²)(day), or 4.7×10^{-4} cm³/(cm²)(sec). If the solute is NaCl, $D_{2s} = 1.6 \times 10^{-5}$ cm²/sec at 25°C, and with $J_1 d/D_{2s} = 1.5$, the feed channel half-height is 0.048 cm, or about 19 mils. With sucrose as solute ($D_{2s} = 0.5 \times 10^{-5}$ cm²/sec), the maximum allowable channel half-height is about 6 mils. Devices with channels this narrow have been constructed, and experiments have been performed in them which are in reasonable agreement with the theory.[107] Maintaining the channel height uniform to prevent flow maldistribution over a path long enough to effect reasonable recovery is an essential requirement. However, an advantage to a thin-channel device is that high membrane packing densities are attainable.

In summary, laminar flow operation offers an alternative to the high viscous-flow pressure drops required to maintain turbulent flow with high flux membranes, but the penalty is increased equipment complexity. Laminar flow operation should be considered when thin channels can be easily devised and when "mixing events" are introduced frequently to maintain entrance region conditions. Polarization effects will generally be large unless recovery between mixing events is small; for seawater desalination, for example, laminar flow is probably impractical. Except for hollow fiber devices, there are no laminar flow reverse-osmosis systems now in practical use, but certain ultrafiltration processes are apparently carried out in the laminar flow regime.

B.   Turbulent Flow

The polarization problem is simpler mathematically in the turbulent flow regime because the flow is fully developed in the first few hydraulic radii and there is no significant entrance region.

Solutions to Equation 26 have been obtained by using film theory. The core of the channel is assumed to be well mixed, and all the resistance to mass transfer is assumed to reside in a stagnant film adjacent to the membrane sometimes called the laminar sublayer. This fictitious layer has an equivalent thickness given by δ ($= \Delta x_s$ in Equation 27). This thickness is determined by the properties of the fluid and the channel dimensions:[2]

$$\delta = \frac{2(D_{2s})^{\frac{1}{3}}\nu^{\frac{2}{3}}N_{\text{Re}}^{\frac{1}{4}}}{0.08\bar{u}_x} \tag{28}$$

where ν is the kinematic viscosity and the Reynolds number, N_{Re}, is given by $4R_h\bar{u}_x/\nu$. Here, R_h is the hydraulic radius defined as the cross-sectional area divided by the wetted perimeter. For a thin channel, the hydraulic radius is just the half-height, d, and

$$\delta = \frac{35.3(D_{2s})^{\frac{1}{3}}\nu^{0.42}\,d^{\frac{1}{4}}}{\bar{u}_x^{\frac{3}{4}}} \tag{29}$$

Combining Equations 27 and 29, we have, for the case of complete solute rejection,

$$\frac{c'_{2s}}{c_{2s}{}^b} = \exp\left[\frac{35.3J_1\nu^{0.42}\,d^{\frac{1}{4}}}{(D_{2s})^{\frac{2}{3}}c_{1s}\bar{u}_x^{\frac{3}{4}}}\right] \tag{30}$$

Several implications are clear from Equation 30. The geometry of the feed channel is of little importance, but water flux, feed velocity, and the solute diffusivity are all important. The essential features have been confirmed experimentally.[103, 108] Most reverse-osmosis units are designed to compromise between frictional losses associated with high feed velocities and concentration polarization and its effect on flux and rejection. For desalination applications the compromise has been based on a solute diffusivity equal to that for sodium chloride ($D_{2s} = 1.6 \times 10^{-5}$ cm²/sec at 25°C), which is representative of the salts in seawater or brackish waters. However, when macromolecules are present, for which $D_{2s} \cong 10^{-7}$–10^{-8} cm²/sec, the polarization modulus can reach enormous values and these molecules gel and become attached to the membrane surface, a phenomenon that leads to one form of fouling. Because solute flux is determined by the concentration in solution at the interface, c'_{2s}, the importance of polarization in a device can be determined by plotting $\ln(J_2/c_{2s}{}^b)$ against $\bar{u}_x^{-\frac{3}{4}}$. The intercept at infinite velocity

$(\bar{u}_x^{-3/4} = 0)$ permits an evaluation of membrane performance free of polarization effects.

An example of the polarization modulus for water desalination in both laminar and turbulent flow is shown in Figure 17.[105] Typical operating conditions of $J_1 = 10$ gal/(ft²)(day) and 50% water recovery were assumed. The channel geometry is a tube with the diameters as indicated, and complete salt

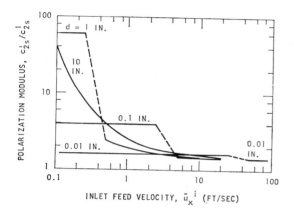

Figure 17.   Polarization modulus vs. inlet feed velocity for laminar and turbulent flow regimes.[105] Average membrane flux = 10 gal/(ft²)(day). Recovery = 50%.

rejection is assumed. A value of $D_{2s} = 1.6 \times 10^{-5}$ cm²/sec was assumed. The polarization modulus used in this figure is not that previously described. Rather, $c_{2s}'/c_{2s}{}^i$ is plotted on the ordinate, where $c_{2s}{}^i$ is the inlet feed concentration.

In Figure 18, the "boundary layer thickness," δ, for NaCl is shown as a function of feed velocity for two channel heights. This graph indicates the reduction in effective channel height resulting from operating in the turbulent regime: effectively very narrow channels are produced. For example, in the asymptotic laminar flow region for a Péclét number of 0.5, a polarization modulus of 1.28 is attained at a channel half-height of 0.017 cm (6.7 mils) for NaCl. The same effective mass transfer coefficient or modulus is attained in turbulent flow in a 1-cm channel at a feed velocity of 14 cm/sec (N_{Re} = 6200). With a 0.1-cm channel, the modulus is always less than this in turbulent flow.

It should be noted that because of uncertainties in membrane performance parameters and channel dimensions in actual reverse-osmosis devices, a certain amount of "tuning" has been practiced in the operation of these units.

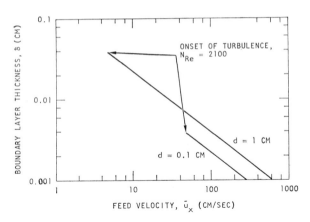

Figure 18. Boundary layer thickness vs. feed velocity.

The effect of feed velocity on performance of the unit is determined empirically,[109] and the operation is modified accordingly. A few data are available on actual system performance in the field. Results for the nominal 5000 gal/day Coalinga pilot plant are probably representative of the performance of a tubular reverse-osmosis unit. This plant contained 1-in.-ID tubes and the average water flux was 20 gal/(ft²)(day). With entry and exit Reynolds numbers of 40,000 and 11,000, respectively, the polarization modulus varied from 1.1 to 1.5 depending on membrane flux and Reynolds number, and the pressure drop through the plant was about 70 psi.[46] In the 50,000 gal/day reverse-osmosis test at the River Valley Golf Course (see Table 5), the average water flux was about 10 gal/(ft²)(day), the polarization modulus was calculated to be about 1.2–1.4, and the system pressure drop was about 40 psi.[47] In this test, the membranes were held 0.11 cm (0.045 in.) apart by a screen which served as a turbulence promoter as well as a brine channel spacer. From the dependence of pressure drop on brine velocity, it was determined that brine flow was in the transition region between laminar and turbulent. Both the Coalinga and River Valley pilot plants were operated at 600 psi and 75% water recovery.

The use of turbulence promoters to enhance mass transfer has been investigated in some detail.[110–112] These devices are commonly used in electrodialysis processing. Recirculation of the feed to increase product recovery and maintain high feed velocities without increasing the membrane area has also been examined. An economic evaluation of these approaches has not been published. The pumping work necessary to overcome frictional losses and the increased osmotic pressure associated with polarization have been calculated for a number of interesting cases;[105] in seawater desalination, excess pumping work can become quite significant.

For anisotropic cellulose acetate membranes, the optimum operating conditions appear to be found in the turbulent flow regime with channel half-heights in the range 0.1–0.5 cm.[105] Long flow channels and, in some instances, series–parallel feed flow arrangements are then required to attain high recovery and maintain high feed velocity. If efficient higher flux membranes are developed, polarization problems will be magnified, and it therefore appears that such membranes might still be operated at the fluxes achievable today but at lower pressures.

ACKNOWLEDGMENT

The author is indebted to the Office of Saline Water for support of some of the studies reported herein, to U. Merten for helpful discussions and for reviewing the manuscript, to R. Riley for his many contributions to this field, and to Gulf General Atomic Incorporated for permission to write this chapter. The assistance of D. Want and G. Hightower in gathering some of the unpublished data reported is appreciated.

APPENDIX. PREPARATION OF ANISOTROPIC CELLULOSE ACETATE MEMBRANES

The original development of the anisotropic membrane has been described in detail by Loeb and Sourirajan.[5] Their procedure was simplified somewhat by Manjikian et al.[42] The preferred procedure can be summarized briefly as follows:

1. The casting solution is a mixture of cellulose acetate, formamide, and acetone, in the proportions 25, 30, and 45% by weight, respectively.
2. The solution is cast at room temperature at a wet thickness of approximately 0.010 in. on an appropriate surface such as a glass plate.
3. After an evaporation period of perhaps 30 sec, the membrane is immersed in ice water, where it is maintained for several minutes while the acetone and formamide are removed.
4. The membrane is annealed in water at 70–95°C for several minutes.

LIST OF SYMBOLS

a Molecular radius
A Membrane constant for water flow
b Coupling coefficient
B Solute permeation constant

c_i	Concentration of component i, mass per unit volume
CRF	Concentration reduction factor $= c'_{2s}/c''_{2s}$
d	Half-height of feed channel
D_i	Diffusion coefficient of component i
f	Frictional coefficient
H	Hydrodynamic permeability
J_i	Diffusive flux of component i
J_v	Volume flux, $\cong J_1$
K	Distribution coefficient, $= c_{2m}/c_{2s}$
L_p	Volumetric permeability parameter, $\cong A$
m	Compaction slope
M	Molecular weight
N	Number of pores per unit area
N_{Re}	Reynolds number
p	Pressure
P	Solute permeability parameter, $\cong B$
r	Pore radius
R	Gas constant
R_h	Hydraulic radius
SR	Solute rejection
t	Time
T	Absolute temperature
u	Velocity of pore fluid
$\bar{u}_x$	Feed velocity averaged over channel height
v_i	Partial molar volume of component i
x	Distance perpendicular to membrane surface
y	Distance down feed channel
Γ	Polarization parameter, $= (c'_{2s}/c_{2s}{}^b) - 1$
δ	Boundary layer thickness in film theory
ε	Porosity
Δx	Effective membrane thickness
ζ	Dimensionless parameter
μ_i	Chemical potential of component i
ν	Kinematic viscosity = viscosity/density
η	Viscosity
π	Osmotic pressure
λ	Tortuosity
σ	Staverman reflection coefficient
τ	Time in days

Subscripts

0	Initial value
1	Water

2	Solute
3	Membrane
d	Value at one day
m	Value in membrane
s	Value in solution

Superscripts

′	Variable evaluated in solution at interface between concentrated solution and membrane
″	Variable evaluated in solution at interface between dilute solution and membrane
b	Variable evaluated in the bulk of the feed solution
i	Inlet value

REFERENCES

1. A. S. Michaels, "Ultrafiltration," in *Progress in Separation and Purification*, Vol. I, E. S. Perry, Ed., Interscience, New York, 1968, p. 297.
2. J. S. Johnson, Jr., L. Dresner, and K. A. Kraus, "Hyperfiltration (Reverse Osmosis)," in *Principles of Desalination*, K. S. Spiegler, Ed., Academic, New York, 1966, Chap. 8, p. 345.
3. E. J. Breton, Jr., "Water and Ion Flow Through Imperfect Osmotic Membranes," *U.S. Off. Saline Water Res. Dev. Prog. Rep.* **16**, University of Florida, 1957.
4. C. E. Reid and E. J. Breton, *J. Appl. Polymer Sci.* **1**, 133 (1959).
5. S. Loeb and S. Sourirajan, "Sea Water Demineralization by Means of a Semipermeable Membrane," *UCLA Water Resources Center Rep. WRCC-34*, 1960; see also *Advances in Chem. Ser.* **38**, 117 (1962).
6. U. Merten, Ed., *Desalination by Reverse Osmosis*, M.I.T. Press, Cambridge, Mass., 1966.
7. U. Merten, "Transport Properties of Osmotic Membranes," in *Desalination by Reverse Osmosis*, U. Merten, Ed., M.I.T. Press, Cambridge, Mass., 1966, p. 15.
8. K. J. Laidler and K. E. Schuler, *J. Chem. Phys.* **17**, 851 (1949).
9. C. E. Rogers, "Solubility and Diffusivity," in *Physics and Chemistry of the Organic Solid State*, D. Fox. M. M. Labes, and A. Weissberger, Eds., Interscience, New York, 1965, pp. 509–635.
10. O. Kedem and A. Katchalsky, *Biochim. Biophys. Acta* **27**, 229 (1958).
11. A. Katchalsky and P. F. Curran, *Nonequilibrium Thermodynamics in Biophysics*, Harvard University Press, Cambridge, Mass., 1967.
12. U. Merten, *Desalination* **6**, 293 (1969).
13. O. Kedem and A. Katchalsky, *J. Gen. Physiol.* **45**, 143 (1961).
14. K. S. Spiegler, *Trans. Faraday Soc.* **54**, 1408 (1958).
15. K. S. Spiegler and O. Kedem, *Desalination* **1**, 311 (1966).
16. G. Scatchard, *J. Phys. Chem.* **68**, 1056 (1964).

17. E. Glueckauf, "On the Mechanism of Osmotic Desalting with Porous Membranes," in *Proceedings of the First International Symposium on Water Desalination, Washington, D.C., October 3–9, 1965*, Vol. 1, U.S. Govt. Printing Office, Washington, D.C., p. 143.

18. K. A. Kraus, R. J. Raridon, and W. H. Baldwin, *J. Am. Chem. Soc.* **86**, 2571 (1964).

19. N. Bjerrum and E. Manegold, *Kolloid-Z.* **43**, 5 (1967).

20. J. D. Ferry, *Chem. Revs.* **18**, 373 (1936).

21. R. A. Robinson and R. H. Stokes, *Electrolyte Solutions*, 2nd Ed., Butterworth, London, 1959.

22. H. Faxén, *Ark. Mat. Astron Fysik* **17** (27) (1922).

23. H. Faxén, *Ann. Physik* **68**, 89 (1922).

24. N. Lakshminarayanaiah, *Chem. Revs.* **65**, 491 (1965).

25. J. R. Pappenheimer, *Physiol. Rev.* **33**, 387 (1953).

26. E. M. Renkin, *J. Gen. Physiol.* **38**, 225 (1954).

27. H. K. Lonsdale, "Properties of Cellulose Acetate Membranes," Chapter 4 in *Desalination by Reverse Osmosis*, U. Merten, Ed., M.I.T. Press, Cambridge, Mass., 1966, p. 93.

28. H. K. Lonsdale, U. Merten, R. L. Riley, and K. D. Vos, "Reverse Osmosis for Water Desalination," *U.S. Off. Saline Water Res. Dev. Prog. Rep.* **150**, General Dynamics, General Atomic Division, 1965.

29. U. Merten, H. K. Lonsdale, R. L. Riley, and K. D. Vos, "Reverse Osmosis for Water Desalination," *U.S. Off. Saline Water Res. Dev. Prog. Rep.* **208**, General Dynamics, General Atomic Division, 1966.

30. U. Merten, H. K. Lonsdale, R. L. Riley, and K. D. Vos, "Reverse Osmosis Membrane Research," *U.S. Off. Saline Water Res. Dev. Prog. Rep.* **265**, General Dynamics, General Atomic Division, 1967.

31. R. L. Riley, C. R. Lyons, and U. Merten, "Transport Properties of Polyvinyl-pyrrolidone-Polyisocyanate Interpolymer Membranes," *Desalination* **8**, 177 (1970).

32. S. Loeb and F. Milstein, *Dechema Monograph.* **47**, 707 (1962).

33. S. Loeb, "Sea Water Demineralization by Means of a Semipermeable Membrane. Progress Report, January 1, 1961 to June 30, 1961," *UCLA Water Resources Center Rep. WRCC-52*, 1962.

34. *Saline Water Conversion Report for 1964*, Office of Saline Water, U.S. Dept. of the Interior, U.S. Govt. Printing Office, Washington, D.C., p. 221.

35. H. K. Lonsdale, U. Merten, and R. L. Riley, *J. Appl. Polymer Sci.* **9**, 1341 (1965).

36. R. L. Riley, J. O. Gardner, and U. Merten, *Science* **143**, 801 (1964).

37. R. L. Riley, U. Merten, and J. O. Gardner, *Desalination* **1**, 30 (1966).

38. R. L. Riley, Gulf General Atomic Inc., private communication.

39. K. D. Vos and F. O. Burris, Jr., *Ind. Eng. Chem. Prod. Res. Dev.* **8**, 84 (1969).

40. P. K. Gantzel and U. Merten, "Gas Separations with High-Flux Cellulose Acetate Membranes," *Ind. Eng. Chem. Proc. Des. Dev.* **9**, 331 (1970).

41. Gulf General Atomic Inc., unpublished data.

42. S. Manjikian, S. Loeb, and J. W. McCutchan, "Improvement in Fabrication Techniques for Reverse Osmosis Desalination Membranes," in *Proceedings of the First International Symposium on Water Desalination, Washington, D.C., October 3–9, 1965*, Vol. 2, U.S. Govt. Printing Office, Washington, D.C., p. 159.

43. S. Manjikian, *Ind. Eng. Chem. Prod. Res. Dev.* **6**, 23 (1967).
44. J. Jagur-Grodzinski and O. Kedem, *Desalination* **1**, 327 (1966).
45. U. Merten, H. K. Lonsdale, R. L. Riley, and K. D. Vos, "Reverse Osmosis Membrane Research," *U.S. Off. Saline Water Res. Dev. Prog. Rep.* **369**, Gulf General Atomic Inc., 1968.
46. J. W. McCutchan and J. S. Johnson, "Reverse Osmosis at Coalinga, California," *J. Am. Water Works Assoc.*, to be published.
47. R. G. Sudak, "Test of 10,000 GPD Reverse Osmosis (Spiral Module) Pilot Plant, San Diego," *U.S. Off. Saline Water Res. Dev. Prog. Rep.* **453**, Gulf General Atomic Inc., 1968.
48. R. G. Sudak, "River Valley Golf Course Test of a Reverse Osmosis Pilot Plant, Part II, 50,000 GPD Capacity," Final Report to Office of Saline Water, U.S. Dept. of the Interior, under Contract 14-01-0001-1264, Gulf General Atomic Inc., to be issued.
49. C. G. de Haven, M. A. Jarvis, and C. R. Wunderlich, "Operation of Reverse Osmosis Pilot Plants," *U.S. Off. Saline Water Res. Dev. Prog. Rep.* **356**, Aerojet-General Corporation, 1968.
50. K. D. Vos, F. O. Burris, Jr., and R. L. Riley, *J. Appl. Polymer Sci.* **10**, 825 (1966).
51. K. D. Vos, A. P. Hatcher, and U. Merten, *Ind. Eng. Chem. Prod. Res. Dev.* **5**, 211 (1966).
52. K. D. Vos, I. Nusbaum, A. P. Hatcher, and F. O. Burris, Jr., *Desalination* **5**, 175 (1968).
53. J. S. Johnson, J. W. McCutchan, and D. N. Bennion, "3-1/2 Years at Coalinga," *UCLA Sch. Eng. Appl. Sci. Rep. No. 69-45*, 1969.
54. J. W. McCutchan, University of California at Los Angeles, private communication.
55. P. A. Cantor, B. J. Mechalas, O. S. Schaeffler, and P. H. Allen, III, "Biological Degradation of Cellulose Acetate Reverse Osmosis Membranes," *U.S. Off. Saline Water Res. Dev. Prog. Rep.* **340**, Aerojet-General Corporation, 1968.
56. R. Blunk, "A Study of Criteria for the Semipermeability of Cellulose Acetate Membranes to Aqueous Solutions." *UCLA Water Resources Center Rep. WRCC-88*, 1964.
57. S. Loeb, "Preparation and Performance of High-Flux Cellulose Acetate Desalination Membranes," Chapter 3 in *Desalination by Reverse Osmosis*, U. Merten, Ed., M.I.T. Press, Cambridge, Mass., 1966. p. 55.
58. H. K. Lonsdale, C. E. Milstead, B. P. Cross, and F. M. Graber, "Study of Rejection of Various Solutes by Reverse Osmosis Membranes," *U.S. Off. Saline Water Res. Dev. Prog. Rep.* **447**, Gulf General Atomic Inc., 1969.
59. H. K. Lonsdale, U. Merten, and M. Tagami, *J. Appl. Polymer Sci.* **11**, 1807 (1967)
60. B. Keilin, "The Mechanism of Desalination by Reverse Osmosis," *U.S. Off. Saline Water Res. Dev. Prog. Rep.* **117**, Aerojet-General Corporation, 1964.
61. R. E. Kesting and J. Eberlin, *J. Appl. Polymer Sci.* **10**, 961 (1966).
62. D. L. Erickson, J. Glater, and J. W. McCutchan, *Ind. Eng. Chem. Prod. Res. Dev.* **5**, 205 (1966).
63. J. P. Agrawal and S. Sourirajan, *Ind. Eng. Chem. Proc. Des. Dev.* **8**, 439 (1969).
64. H. K. Lonsdale, R. L. Riley, C. E. Milstead, L. D. LaGrange, A. S. Douglas, and S. B. Sachs, "Research on Improved Reverse Osmosis Membranes," *U.S. Off. Saline Water Res. Dev. Prog. Rep.* **577**, Gulf General Atomic Inc., 1970.

65. S. Loeb and S. Manjikian, "Brackish Water Desalination by an Osmotic Membrane," *UCLA Water Resources Center Rep. WRCC-78*, 1963.

66. A. S. Michaels, H. J. Bixler, and R. M. Hodges, Jr., *J. Colloid Sci.* **20**, 1054 (1965).

67. R. E. Kesting, W. J. Subcasky, and J. D. Paton, *J. Colloid Interface Sci.* **28**, 156 (1968).

68. J. Schultz, A. Riedinger, and H. McCracken, "Brackish Well Water Reverse Osmosis Tests at Midland, Ft. Stockton, and Kermit, Texas," *U.S. Off. Saline Water Res. Dev. Prog. Rep.* **237**, General Dynamics, General Atomic Division, 1966.

69. C. W. Saltonstall, Jr., F. C. Burnette, W. S. Higley, W. M. King, and A. L. Vincent, "A Study of Membranes for Desalination by Reverse Osmosis," *U.S. Off. Saline Water Res. Dev. Prog. Rep.* **232**, Aerojet-General Corporation, 1966.

70. C. W. Saltonstall, Jr., "Development and Testing of High-Retention Reverse Osmosis Membranes," *International Conference PURAQUA, Rome, Italy. February 17–22. 1969*, to be published.

71. D. L. Hoernschemeyer, C. W. Saltonstall, Jr., O. S. Schaeffler, L. W. Schoellenbach, A. J. Secchi, and A. L. Vincent, "Research and Development of New and Improved Cellulose Ester Membranes," *U.S. Off. Saline Water Res. Dev. Prog. Rep.* **556**, Aerojet-General Corporation, 1970.

72. S. Manjikian, S. Liu, M. Foley, C. Allen, and B. Fabrick, "Development of Reverse Osmosis Membranes," *U.S. Off. Saline Water Res. Dev. Prog. Rep.* **534**, Universal Water Corporation, 1970.

73. R. L. Riley, H. K. Lonsdale, C. R. Lyons, and U. Merten, *J. Appl. Polymer Sci.* **11**, 2143 (1967).

74. R. L. Riley, H. K. Lonsdale, L. D. LaGrange, and C. R. Lyons, "Development of Ultrathin Membranes," *U.S. Off. Saline Water Res. Dev. Prog. Rep.* **386**, Gulf General Atomic Incorporated, 1968.

75. W. P. Cooke, "Performance of Hollow Fiber 'Permasep' Permeators in Industrial Water and Waste Stream Purification and Separations," *International Conference PURAQUA, Rome, Italy, February 17–22, 1969.*

76. R. J. Mattson and V. J. Tomsic, *Chem. Eng. Prog.* **65**, 62 (1969).

77. W. E. Skiens and H. I. Mahon, *J. Appl. Polymer Sci.* **7**, 1549 (1963).

78. H. I. Mahon, U.S. Pat. 3,228,876 (1966).

79. H. I. Mahon, E. A. McLain, W. E. Skiens, B. J. Green, and T. E. Davis, *Chem. Eng. Prog. Symp. Ser.* **65** (91), 48 (1969).

80. F. Gotch, B. Lipps, J. Weaver, Jr., J. Brandes, J. Rosen, J. Sargent, and P. Oja, *Trans. Am. Soc. Artif. Intern. Organs* **XV**, 87 (1969).

81. J. G. McKelvey, Jr., K. S. Spiegler, and M. R. J. Wyllie, *Chem. Eng. Prog. Symp. Ser.* **55** (24), 199 (1959).

82. F. G. Donnan, *Z. Electrochem.* **17**, 575 (1911); see also F. Helfferich, *Ion Exchange*, McGraw-Hill, New York, 1962.

83. *Saline Water Conversion Report for 1969*, Office of Saline Water, U.S. Dept. of the Interior, U.S. Govt. Printing Office, Washington, D.C.

84. A. E. Marcinkowsky, K. A. Kraus, H. O. Phillips, J. S. Johnson, Jr., and A. J. Shor, *J. Am. Chem. Soc.* **88**, 5744 (1966).

85. K. A. Kraus, H. O. Phillips, A. E. Marcinkowsky, and J. S. Johnson, *Desalination* **1**, 225 (1966).

86. K. A. Kraus, A. J. Shor, and J. S. Johnson, Jr., *Desalination* **2**, 243 (1967).

87. A. J. Shor, K. A. Kraus, W. T. Smith, Jr., and J. S. Johnson, Jr., *J. Phys. Chem.* **72**, 2200 (1968).

88. J. R. Kuppers, A. E. Marcinkowsky, K. A. Kraus, and J. S. Johnson, *Separation Science* **2** (5), 617 (1967).

89. L. C. Flowers, D. E. Sestrich, and D. A. Berg, "Reverse Osmosis Membranes Containing Graphitic Oxide," *U.S. Off. Saline Water Res. Dev. Prog. Rep.* **224**, Westinghouse Electric Corporation, 1966.

90. F. E. Litman and G. A. Guter, "Research on Porous Glass Membranes for Reverse Osmosis," *U.S. Off. Saline Water Res. Dev. Prog. Rep.* **379**, McDonnell Douglas Corporation, 1968.

91. H. Bechhold, *Z. Physik. Chem.* **60**, 257 (1907).

92. W. J. Elford, *Proc. Roy. Soc. (London)* B **112**, 384 (1933).

93. R. Collander, *Soc. Sci. Fenn. Commentat. Biol.* [6] **2**, 1 (1926).

94. J. W. McBain and W. L. McClatchie, *J. Am. Chem. Soc.* **55**, 1315 (1933).

95. R. P. Durbin, *J. Gen. Physiol.* **44**, 315 (1960).

96. B. Z. Ginzburg and A. Katchalsky, *J. Gen. Physiol.* **47**, 403 (1963).

97. W. E. Henderson and C. M. Sliepcevich, *Chem. Eng. Prog. Symp. Ser.* **55** (24), 145 (1959).

98. A. S. Michaels, *Chem. Eng. Prog.* **64**, 31 (1968).

99. A. S. Michaels, *Ind. Eng. Chem.* **57**, 32 (1965).

100. A. S. Michaels, U.S. Pat. 3,276,598 (1967).

101. "Pellicon Ultrafiltration," *Millipore Corporation Application Rep. AR-21*, 1969.

102. L. Nelson, "The Design and Testing of a Narrow Channel Laminar Flow Hemodiafilter (The Diaphron), Second Year Report," *Ref. No. PH 43-66-45*, Amicon Corporation, 1969.

103. T. K. Sherwood, P. L. T. Brian, R. E. Fisher, and L. Dresner, *Ind. Eng. Chem. Fundam.* **4**, 113 (1965).

104. P. T. L. Brian, *Ind. Eng. Chem. Fundam.* **4**, 439 (1965); see also *M.I.T. Desalination Res. Lab. Rep. 295-7*, 1965.

105. P. L. T. Brian, "Mass Transport In Reverse Osmosis." Chapter 5 in *Desalination by Reverse Osmosis*, U. Merten, Ed., M.I.T. Press, Cambridge, Mass., 1966, p. 161.

106. W. N. Gill, C. Tien, and D. W. Zeh, *Ind. Eng. Chem. Fundam.* **4**, 433 (1965).

107. H. Strathmann, "Control of Concentration Polarization in Reverse Osmosis Desalination of Water," *U.S. Off. Saline Water Res. Dev. Prog. Rep.* **336**, Amicon Corporation, 1968.

108. U. Merten, H. K. Lonsdale, and R. L. Riley, *Ind. Eng. Chem. Fundam.* **3**, 210 (1964).

109. D. T. Bray, "Engineering of Reverse-Osmosis Plants," Chapter 6 in *Desalination by Reverse Osmosis*, U. Merten, Ed., M.I.T. Press, Cambridge, Mass., 1966, p. 203.

110. D. G. Thomas, *A.I.Ch.E.J.* **12**, 124 (1966).

111. J. S. Watson and D. G. Thomas, *A.I.Ch.E.J.* **13**, 676 (1967).

112. D. G. Thomas and J. S. Watson, *Ind. Eng. Chem. Proc. Des. Dev.* **7**, 397 (1968).

BIBLIOGRAPHY

H. K. Lonsdale, "Separation and Purification by Reverse Osmosis," in *Progress in Separation and Purification*, Vol. III, E. S. Perry and C. J. Van Oss, Eds., Wiley, New York, to be published.

U. Merten, Ed., *Desalination by Reverse Osmosis*, M.I.T. Press, Cambridge, Mass., 1966.

A. S. Michaels, "Ultrafiltration," in *Progress in Separation and Purification*, Vol. I, E. S. Perry, Ed., Wiley, New York, 1968, p. 297.

Proceedings of the First International Symposium on Water Desalination, Washington, D.C., October 3–9, 1965, U.S. Govt. Printing Office, Washington, D.C., 1967.

Saline Water Conversion Reports, U.S. Dept. of the Interior, U.S. Govt. Printing Office, Washington, D.C.

K. S. Spiegler, Ed., *Principles of Desalination*, Academic, New York, 1966.

Chapter IX   The Costs of Reverse Osmosis

Robert E. Lacey*

Although few details have been published about the costs of industrial processing by reverse osmosis, many reports have been issued on the capital and operating costs of treating brackish water or seawater by reverse osmosis.[1-7] Because several aspects of the technical and costing practices associated with desalting water differ from the practices normally used in industry (e.g., amortization schedules, interest rates, permeate-to-concentrate ratios), the costs and optimization of costs given in the reports on desalting water are not directly applicable to industrial processing with reverse osmosis. Nevertheless, the reports on desalination costs afford valuable information about the effects of various operating conditions on costs.

Because of the continuing rapid development of reverse-osmosis membranes and equipment it is difficult to present a precise economic analysis of reverse-osmosis processing with any confidence that the analysis will be of value in the future. A major cause of the rapid changes in reverse-osmosis technology is that there are four basic types of reverse-osmosis equipment (i.e., plate-and-frame, tubular, spiral-wound, and hollow-fiber) in competition with one another. In addition, several types of membrane materials may be competitors (see Chapter 8). The membrane area required for a given production rate varies with the type of equipment used. The percentage of the feed solution recovered as permeate and the concentration of solute in the permeate are affected strongly by the type of membranes used. In addition, a number of studies are in progress by both the Office of Saline Water and private industry to develop improved membrane materials and equipment.

---

* Southern Research Institute, Birmingham, Alabama.

Because of these rapid developments in membranes and equipment, it is more worthwhile to present general cost information that will be of lasting value than to give detailed cost analyses.

I. COST EQUATIONS AND OPTIMIZATION OF COSTS

Among the operating conditions that affect the capital and operating costs of reverse-osmosis processing are the following:

- The desired rate of production.
- The nature and composition of the feed solution.
- The nature and composition of the desired product (which may be either the permeate or the concentrated feed solution).
- The flux of permeate per unit pressure drop for the type of membrane used and the nature of the permeate.
- The flux of solute through the type of membranes used.
- The applied pressure.
- The type of reverse-osmosis equipment used.
- The ratio of permeate flow to the flow of high-pressure concentrate.

Lonsdale et al.[3] developed an equation that describes the contributions of the amortization of the capital costs of (a) the high-pressure pumps, (b) the membrane permeator equipment, and (c) the circulation pumps relative to the total cost of producing a unit weight of permeate as functions of the differential pressure (ΔP) and of the solute concentration in the concentrated high-pressure stream (C_B). The equations also include terms that describe the contribution of the cost of electrical energy (for pumping) to the total processing cost of a unit of permeate in terms of ΔP and C_B. With these equations the contributions of the amortization of the capital cost of the various portions of a reverse-osmosis plant and of the cost of pumping energy can be calculated for various assumed values of ΔP and C_B, to show the variations in the individual factors contributing to total processing costs with changes in ΔP and C_B. Alternatively, the partial differentials of the cost equations with respect to ΔP and to C_B can be equated to zero and the values of ΔP and C_B at which the lowest processing cost is obtained can be calculated. This approach to optimizing the cost of processing by reverse osmosis is valuable in that it not only identifies the values of operating conditions that result in low costs, but it also shows which cost factors contribute most to total processing cost and how these cost factors vary with changes in ΔP and C_B.

Menzel[5] and Bray[7] refined the approach of Lonsdale et al. and included (a) membrane-replacement cost, (b) operational and maintenance costs, and

(c) the flux of permeate per unit pressure drop as terms in the cost equation in addition to the terms Lonsdale used. In Menzel's equations the total product water cost, C_t, was calculated by the following general equation.

$$C_t = C_{pp} + C_m + C_{o+a} + C_e + C_{mr} + C_{t+i} + C_{iwc},$$

where individual costs were defined as follows:

C_{pp} = cost of pumping power (based on 0.7 ¢/kW-hr)

C_m = cost of materials, including feed water, chemicals, supplies, and maintenance materials. The feed-water cost was assumed to be nil. The cost of feed-water treatment was assumed to be 1.5¢/1000 gal of feed water.* (Costs for supplies and maintenance were assumed to be 0.5% per annum of total plant investment.)

C_{o+a} = cost of operation and administration personnel, including operational and maintenance labor, payroll extras, and general and administrative overhead. The total time for operating and maintenance labor was assumed to be 4888 hr/yr,† with an hourly rate of $3.50. For payroll extras, 15% of the operating and maintenance charges was added. To cover general and maintenance overhead, the total costs for operating and maintenance labor and the payroll extras were multiplied by a factor of 1.3.

C_e = amortized equipment cost, including land, building, etc. Based on a 20-yr plant life and 4% interest, the amortization rate is 7.4% per annum.

C_{mr} = cost of membrane replacement included as a parameter and varied from $0.10 to 1.00/(ft²)(yr)

C_{t+i} = cost of taxes and insurance, based on 2% per annum of total investment

C_{iwc} = cost of interest on working capital, assumed to be 4% per annum

Menzel then developed expressions for each individual cost (e.g., C_{pp}, C_m) in terms of appropriate operating variables, such as applied pressure, salt rejection, and permeate-to-brine flow ratios. His expressions for the individual costs are given in the original reference. With the use of a computer, Menzel could easily determine the effects of changes in operating variables on total costs and on individual costs.

* Assuming use of 1.04 lb of H_2SO_4 (at $25/ton) per 1000 gal of feed water and a cost for chlorine of 0.2¢/1000 gal of feed water.
† Based on 1 man × 8 hr/day for 365 days = 2920 hr, 1 man × 8 hr/day for 14 days = 112 hr (vacation replacement), 2 men × 8 hr/day for 30 days = 480 hr (maintenance and membrane replacement), 1 man × 8 hr/day for 172 days = 1376 hr (equivalent to ½ yr supervision at $9600/yr salary).

The costs of land and building and of engineering are essentially constant, and not affected by either the system pressure or the flow ratio. The cost of the feed-pump subsystem and the filter subsystem are dependent on the flow ratio, and the cost of the main-pump subsystem and the membrane–pressure vessel subsystem are dependent on both the flow ratio and the system pressure.

The amortization schedules and interest rates assumed by Menzel for the desalination costs he developed (20-yr life with interest at 4%) are not applicable to industrial processing, and the membrane replacement costs he assumed ($0.50/ft² of membrane surface per year) might be achieved only in large plants (such as desalination plants) in which membrane replacement might be partly or entirely automated. Nevertheless, the studies of Menzel and Bray afford information of value to industrial processing, and the costing procedures described by Menzel are of value as a guide for the engineer in industry, who can use amortization schedules, interest rates, unit costs of energy, and membrane replacement costs that are appropriate to industrial practice instead of the rates and costs used by Menzel.

II. EFFECTS OF OPERATING VARIABLES ON CAPITAL AND OPERATING COSTS

For a given throughput of permeate, Figures 1 and 2* show the effects on capital cost of variations in the system pressure and the ratio of the flow of permeate to that of the high-pressure concentrate. The curves in Figure 1 were developed assuming the use of a membrane with moderate permeability (nominal membrane constant = 1×10^{-5} g/(sec)(cm²)(atm)), and those in Figure 2 were developed assuming the use of a membrane with twice that permeability.

Because of the interaction between the costs that are dependent on the flow ratio and those that are dependent on the system pressure, and because the flux of permeate depends on system pressure, there is an optimum flow ratio for each pressure. Conversely, there is an optimum pressure for any given flow ratio. Comparison of Figures 1 and 2 shows that the variations of capital cost with system pressure and flow ratio are less with highly permeable membranes (Figure 2) than with moderately permeable membranes. The comparison also shows that the capital costs for a given throughput of permeate are lower with highly permeable membranes than with moderately permeable ones.

Figures 3 and 4 show the variations in total processing cost per unit of permeate (including operating and maintenance costs and the cost of

* The curves in Figures 1–6 are based on data given by Menzel.[5]

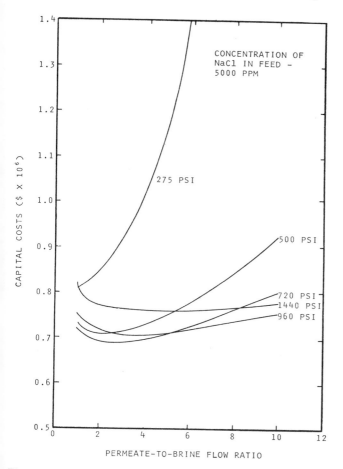

Figure 1. Capital costs for a reverse-osmosis plant to produce 10^6 gal/day of permeate as a function of applied pressure and permeate-to-brine flow ratio (nominal membrane constant $= 1 \times 10^{-5}$ g/(sec)(cm^2)(atm)).

amortizing the capital investment) with changes in system pressure and flow ratio. Figure 3 is for a system with moderately permeable membranes and Figure 4 is for a system with highly permeable membranes.

With both membranes, there are optimum system pressures and flow ratios to achieve the lowest cost of permeate, just as there are to achieve the lowest capital cost. Comparison of Figures 3 and 4 shows that the cost of processing is lower with highly permeable membranes than with moderately permeable ones.

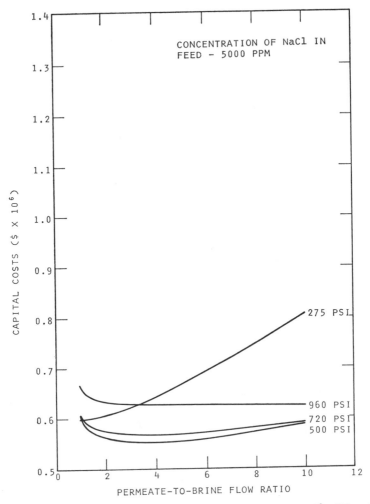

Figure 2.   Capital costs for a reverse-osmosis plant to produce 10^6 gal/day of permeate as a function of applied pressure and permeate-to-brine flow ratio (nominal membrane constant = 2×10^{-5} g/(sec)(cm²)(atm)).

Figures 5 and 6 show the relationships between cost and permeate-to-brine ratio for the individual cost factors that contribute to the total processing cost with moderately permeable membranes (Figure 5) and highly permeable ones (Figure 6). A pressure near the optimum for each system was chosen to develop the data in Figures 5 and 6—960 psi for Figure 5 and 720 psi for Figure 6.

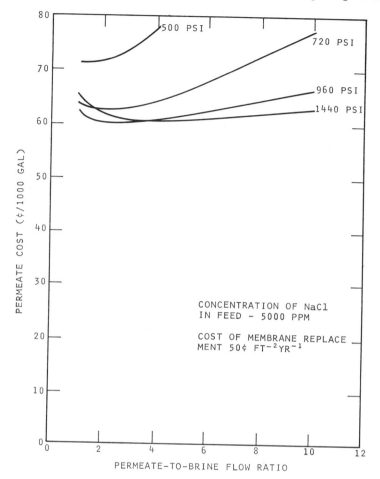

Figure 3. Cost of permeate from a reverse-osmosis plant that produces 10^6 gal/day as a function of applied pressure and permeate-to-brine flow ratio (nominal membrane constant $= 1 \times 10^{-5}$ g/(sec)(cm^2)(atm)).

The membrane replacement cost and annual capital charges are the most important individual costs for systems with either type of membrane. If industrial practice in amortizing the capital costs were followed instead of the practices for desalination, the annual capital charges would be the largest factor in costs for systems with either type of membrane. The increase in membrane replacement cost with increase in permeate-to-brine flow ratios shown in Figures 5 and 6 results from increases in back-osmotic pressure

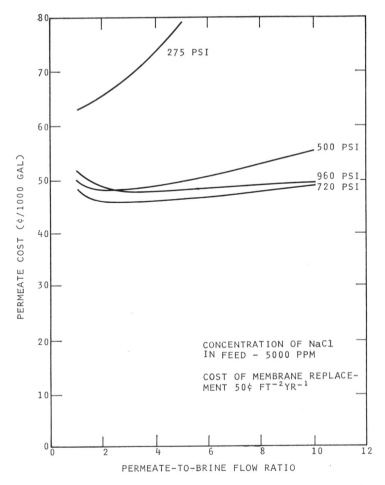

Figure 4. Cost of permeate from reverse-osmosis plant that produces 10^6 gal/day as a function of applied pressure and permeate-to-brine flow ratio (nominal membrane constant = 2×10^{-5} g/(sec)(cm²)(atm)).

since the brine stream becomes more concentrated as the flow ratio increases. Comparison of Figures 5 and 6 shows that the membrane replacement cost for the system with highly permeable membranes (Figure 6) is considerably lower than that for the system with moderately permeable membranes, as would be expected since less membrane area is required. The lower annual capital charges for the system with highly permeable membranes is attributable to both the lower system pressure (720 psi compared with 960 psi) and the lower membrane area required.

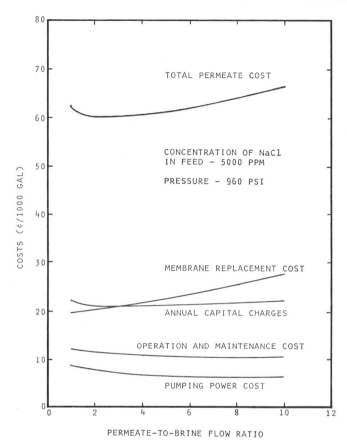

Figure 5. Contributions of individual cost factors to the total cost of permeate from a reverse-osmosis plant that produces 10^6 gal/day as a function of permeate-to-brine flow ratio (nominal membrane constant $= 1 \times 10^{-5}$ g/(sec)(cm^2)(atm)).

No attempt should be made to apply the absolute values of processing cost shown in Figures 3–6 to a particular industrial separation problem because of the differences between desalination and industrial costing practices. The costs for industrial separations will usually be somewhat higher than those indicated. Even though the costs are not directly applicable to industrial practice, the relationships between operating variables and costs shown in Figures 1–6 show general variations of costs with operating variables that should be applicable to industrial problems.

In addition, the equations developed by Menzel and Bray can be used by the engineer in industry with appropriate changes (e.g., in amortization times,

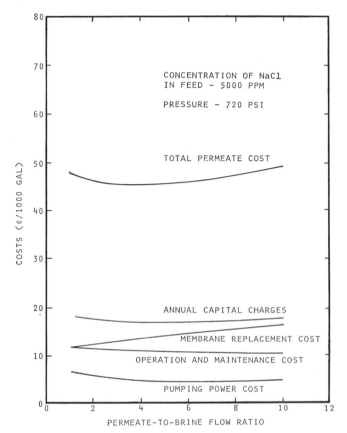

Figure 6. Itemized cost of permeate from a reverse-osmosis plant that produces 10^6 gal/day as a function of permeate-to-brine flow ratio (nominal membrane constant = 2×10^{-5} g/(sec)(cm^2)(atm)).

interest rates, and costs of labor) to compute the costs for an industrial separation.

REFERENCES

1. S. Loeb and S. Sourirajan, *Advan. Chem. Ser.* **38,** 117 (1963).
2. B. Keilen and C. G. de Haven, "Design Criteria for Reverse Osmosis Desalination Plants," *Proceedings of the International Symposium on Water Desalination, October 1965, Washington, D.C.*
3. H. K. Lonsdale, U. Merten, R. L. Riley, K. D. Vos, and J. C. Westmoreland, *U.S. Off. Saline Water Res. Dev. Rep.* **111** (1964).

4. L. T. Fan, C.Y Cheng, L. Y. S. Ho, C. L. Hwang, and L. Erickson, *Eng. Exp. Station Spec. Rep.* **73**, Kansas State University, Manhattan, Kansas (1967).
5. H. F. Menzel, *U.S. Off. Saline Water Res. Dev. Rep.* **236** (1967).
6. W. L. Prehn and J. L. McGaugh, *U.S. Off. Saline Water Res. Dev. Rep.* **555** (1970).
7. D. T. Bray, "Engineering of Reverse Osmosis Plants," Chapter 6 in *Desalination by Reverse Osmosis*, U. Merten, Ed., M.I.T. Press, Cambridge, Mass., 1969.

Chapter X   Reverse Osmosis
in the Food Industry

R. L. Merson* and L. F. Ginnette†

*The Lord God took the man and put him in
the garden of Eden to till it and keep it.*
Genesis 2: 15

I. FOOD TECHNOLOGY AND THE NEED FOR REVERSE OSMOSIS

From antiquity men have sought improved methods of producing and preserving adequate supplies of food. In some parts of the world the search has been eminently successful, and several societies now enjoy an abundance of food. As a result, a large proportion of the population in these so-called "developed" countries has been freed of any necessity to engage in food production and now performs other work, usually in urban communities. In the United States, the example *par excellence*, farm population dwindled from 35% of the total population in 1910 to 5% in 1968. This dramatic urbanization has been accompanied by the development of huge industries devoted to food processing, preservation, and transportation. Over the years, a number of techniques have been devised for improving the quality and safety of preserved food and the efficiency with which it is brought to market. Canning, pickling, smoking, freezing, and fermentation are familiar examples.

* Department of Food Science and Technology, University of California, Davis.
† Western Regional Research Laboratory, Agricultural Research Service, U.S. Department of Agriculture, Albany, California.

A. Methods of Food Preservation

Drying is another important technique which has advantages both for preservation and efficient transportation. Almost all living organisms that compose our source of food contain more water than anything else. The fluids in and around the living cells provide a mobile medium in which the life processes occur. After the normal life of the organism has been interrupted (the fruit plucked from the tree or the animal slaughtered) biological and chemical changes may continue in the moist environment, and the organism may serve as a growth medium for microbial flora. Food preservation normally involves establishing conditions that will prevent or drastically slow the rate of undesirable changes. The earliest forms of food preservation, for example, salting of fish or drying of meat and fruit, consisted of reducing the activity of water in the food to the point that these changes could occur only very slowly. Drying is still today the most prevalent method of food preservation.

In 1810 in France, Nicolas Appert introduced a radically new method of food preservation, *Appertizing* (now called canning), which consisted of heating foods in sealed containers. Canning deactivates microorganisms and enzymes in the food by heat and the sterile product is protected from recontamination by the container. This method eliminates the expense of drying the food but often drastically changes its taste and texture, and introduces the expense of the protective package.

In the present century we have become interested not only in safe preservation but also in preserving the nutritive value and the original flavor. Preservation by freezing and cold storage was a major advance in retaining the fresh qualities of foods that are not ordinarily cooked. Increased convenience is also a goal of modern food processing, convenience in terms of year-round availability, lower costs, and ease of reprocessing, distribution, storage, and use.

B. Commercial Dewatering of Foods

Dewatering, including concentration (partial dehydration), of liquid foods plays a major role in each method of preservation: in drying, in canning, and in freezing of foods. Selective removal of water results in less material and hence lower costs for heating, canning, freezing, packaging, transporting, and storing. At present, dewatering is achieved commercially by atmospheric or vacuum evaporation. The final concentrate from the evaporator usually has an osmotic pressure of 1000–1200 psi (42° Brix orange juice concentrate, 33% tomato paste, 36% solids milk concentrate, etc.). In special cases water is removed by other methods, by freezing out pure water crystals, or by distillation, as in the alcoholic beverage industry where the product is more volatile than water.

Dewatering is already a well-established and sizable industry. A few examples of commodities that are concentrated or dried are listed in Table 1. Many others could be cited, such as instant coffee, soups, cheese whey, and maple, corn, and cane sugar syrups. However, large quantities of liquid food are still sold in their natural state because of the higher quality of this fresh

Table 1. Commercial Dewatering of Foods, Approximate Commercial Volume in 1967[7, 50]

Commodity	Initial Solids Content (%)	Product	Product Solids Content (%)	Water Removed ($10^6 \times$ gal/yr)
Tomato juice	4–6	Catsup	18	330
		Paste	30	440
Orange juice	10–12	Concentrate	45	340
Milk	12	Concentrate	24	220
Beet sugar	15	Solid	100	5000

food and the high cost of processing. There is a vast potential market as yet untapped, if processing costs can be lowered, if the quality of the dewatered products can be improved, or if new processing methods can be devised for commodities for which no suitable processing exists. There is need for a gentle, economical process such as reverse osmosis to meet these objectives.

II. ADVANTAGES OF REVERSE OSMOSIS

In concentrating foods, whether by reverse osmosis or some more traditional method, the ideal objective is to remove only water (and very occasionally some other undesired component) while retaining as much as possible of the original flavor, aroma, nutritional value, and functional properties of the food. Actually, it is generally not possible to retain completely all desirable properties, in spite of the enthusiastic claims that have been made for one or another novel method, such as freeze-drying or freeze-concentration. Reverse osmosis is no exception to the rule that all processing involves some loss of quality. Nevertheless, since it involves neither elevated temperatures nor evaporation, it may possess some unique advantages.

A. Retention of Flavor and Aroma

The processes by which foods acquire their characteristic flavor and aroma are not really well understood, although the tempo of flavor research is increasing and much progress will undoubtedly be made in the next decade. It is known, however, that most foods (fruit juices, for example) contain small amounts of volatile organic substances which contribute in some way to the identity and intensity of their flavor and aroma. In many cases, hundreds of different molecular species have been detected in the vapors over a single sample of a food material. These volatile substances tend (as a class) to possess a very high activity in dilute water solutions. For this reason they appear in relatively high concentration in the vapor given off by foods and are stripped out and lost during evaporative concentration or drying. On the other hand, these volatile substances tend to be rejected by selective membranes, although practical membranes capable of quantitatively rejecting all volatile molecules except water have not yet appeared.

B. Preservation of Functional Properties

Besides retaining volatile flavor and aroma components, it is sometimes necessary to preserve some functional property, for example, the foam-stabilizing (whipping) property of egg white. This is often very difficult to do by ordinary evaporation or dehydration methods. Commercial dehydration of egg white without severely damaging its whipping characteristics can be done at present only by means of freeze-drying. (Other forms of dehydrated egg white contain additives to restore the whipping property.) Egg white can, however, be concentrated by reverse osmosis without loss of functional properties, and without requiring subsequent addition of a whipping agent.

C. Economic Advantages

One expects reverse osmosis to have an inherent economic advantage over evaporation or freeze-concentration because no phase change is required of the water. The latter methods require additional costs for energy to change liquid water to vapor or to ice, energy which is never completely recovered. Also, since food raw materials are produced over a large geographical area, they have been processed traditionally in small plants rather than in a central location. Thus it is particularly important that the economics and efficiency of reverse osmosis are not markedly dependent on the size of the plant. Low capital investment is also a major factor to an industry which historically expects a return on its investment with a single crop or growing season.

III. FOOD APPLICATIONS OF REVERSE OSMOSIS

We have seen that reverse osmosis can be expected to compete with other methods of water removal when it has an economic advantage, when the fragility of the product requires gentle treatment, or when fractionation other than dewatering (e.g., salt removal) is desired. A number of applications that have been studied are listed in Table 2.

Commercially, cheese and cottage cheese wheys are being processed in a number of plants in the United States, Australia, and New Zealand. Whey processing usually involves separating the solutes into fractions containing

Table 2.   Applications of Reverse Osmosis in the Food Industry

---
Permeate is of Major Interest
  Recovery of water from waste streams[49b]
  Recovery of brines (pickles, olives, etc.) by passing salts, acids, and water through membrane
  Permeating food-grade lactose from solutions of whey protein and lactose[16, 29, 49a]
Concentrate is of Major Interest
  Concentrating fruit juices,[10, 34] coffee, and maple sap[48, 49d, 51] without phase change, heat damage, or loss of volatiles[39]
  Concentrating protein,[5, 6] egg white,[27] whey protein,[16, 49a] and gelatin without heat or shear damage to functional properties
  Concentrating pectin solutions without degrading the chain length (and hence sugar-supporting power) of the pectin molecules
  Concentrating egg white, pineapple mill juice, and sugar solutions[45] without browning reactions
  Deashing beet sugar solutions by passing salts and water through the membrane[52]
  Desugaring egg white by passing glucose through the membrane
  Purifying whey protein and lactose by passing lactic acid and salts[30, 40, 49a]

---

mostly protein, lactose, or salt and lactic acid as well as concentrating each of these fractions. A large reverse-osmosis plant in California removes solutes from the effluent from a food fermentation process. Hydronautics-Israel, Ltd., has operated a two-stage reverse-osmosis pilot plant for concentrating orange juice.[10]

The applications of reverse osmosis all depend, of course, on the ability of the membrane to retain desirable food solutes with the concentrated product. Indeed in many food applications, as we shall see later, the *rate* of water permeation is not limited by the membrane itself but rather by slow mass transport in the liquid food phase. The function of the membrane then is to provide solute rejection. In the next section we examine what is presently known about transport of food solutes through membranes.

IV. PERMEABILITY OF MEMBRANES TO FOOD SOLUTES

The constituents of foods cover an extremely broad range of physical and chemical characteristics. Liquid foods contain insoluble solids and solutes which are infinitely soluble in water; they contain solutes such as sugars up to 20% of their weight and solutes in trace quantities too small to measure by present methods; they contain some of the simplest and the most complex compounds known to man.

One class of compounds, the aroma compounds, deserves special note. As in Figure 1, these compounds can be grouped into oil-soluble compounds

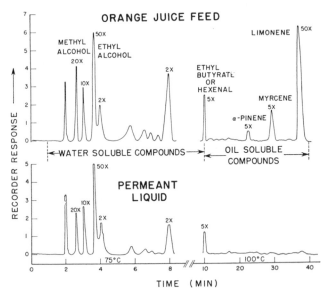

Figure 1. Chromatographic representation of the volatile aroma compounds of orange juice. Havens Module Type 5A operated at 1500 psi at approximately 5C.

(high-molecular-weight and/or nonpolar molecules) which do not permeate the membrane and water-soluble species (low-molecular-weight, polar compounds) which do. Members of the latter group exhibit very interesting behavior in reverse osmosis. They are often chemically similar to water (high oxygen and hydroxyl content), are present in foods in minute quantities (concentrations on the order of 10 ppb to 10 ppm), and have such high activity, even in dilute solution, that rejection by the membrane is poor except for very tight membranes. Selectivity data[31] for homologous series of straight-chain alcohols and for a series of normal esters of acetic acid are shown in Figures 2 and 3. These data were obtained at 1000 psi for 0.001 M solutions of these compounds with three different cellulose acetate membranes. A short channel[34] and high flow rates were used to minimize concentration

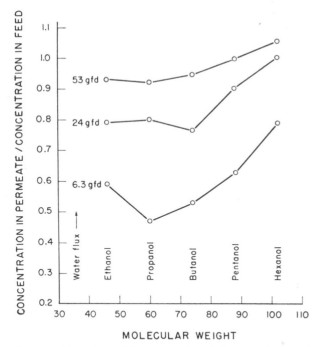

Figure 2. Retention of low-molecular-weight normal alcohols by Loeb-type cellulose acetate membranes.[34] Feed concentration of alcohols, 100 ppm; operating pressure, 1000 psi; temperature, 20°C.

polarization. However, concentrations were measured by gas–liquid chromatography by direct liquid injection with a precision of only about ±10%. Note that in the alcohol series propanol to hexanol, rejection is poor and becomes progressively worse with increasing molecular weight. This is in contradiction to statements in the literature[35] which point out that rejection usually increases with molecular weight in an homologous series. However, Sourirajan[44] has noted that propanol is rejected better than either ethanol or butanol for more concentrated solutions (0.5 M). The ester series (Figure 3) shows a minimum rejection with butyl acetate, whereas the rejection would be expected to improve with increasing molecular weight. The apparent negative rejection of hexanol with the high flux membrane agrees with results[25] for phenol, another six-carbon alcohol. However the analytical procedures for the very dilute solutions used here are not sufficiently accurate to state categorically that negative rejection is occurring.

That molecular structure is a very important factor in membrane rejection is also evident in Figure 4. Note especially the curves for ethyl butyrate and

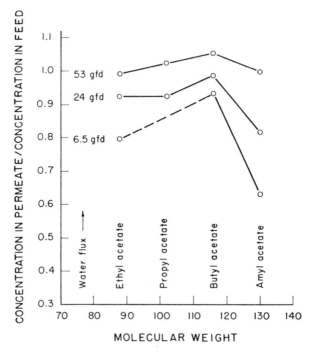

Figure 3. Retention of normal esters of acetic acid. Temperature 20–24°C. Other conditions as in Figure 2 and Reference 34.

dextrose, both of which are six-carbon molecules. In general, the monosaccharide and disaccharide food sugars (MW 180–360) are rejected ($>95\%$) by cellulose acetate membranes with pure water permeation rates below 30–40 gal/(ft²)(day).[20, 34] Food acids, citric (MW 192), malic (MW 134), lactic (MW 90), and acetic (MW 60) show decreasing rejections; acetic acid is a solvent for the cellulose diacetate. Salts are sometimes difficult to retain if a high flux membrane is used but this is no serious problem in food processing and is often an advantage, as in desalting of whey.

Insoluble solids, bacteria, enzymes (MW 50,000–200,000), proteins (MW 50,000–100,000), and other very large molecules are rejected by even very porous ultrafiltration membranes (intrinsic water rates of 65 gal/(ft²)(day) at 10 psi).[1, 8, 37] Polysaccharides (MW ~20,000) require a more selective membrane but are completely retained by membranes with rates up to 90 gal/(ft²)(day) at 1000 psi[1] or 175 gal/(ft²)(day) at 100 psi.[2, 36–38]

Most of the separation data for foods have been determined for components in situ; the whole liquid food has been fed into a reverse-osmosis unit and analysis of the inlet, outlet, and permeate streams has yielded

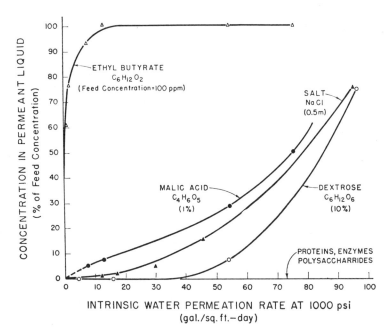

Figure 4. Retention of food solutes by cellulose acetate membranes.[33] Feed concentrations are shown in parentheses.

rejection data. These are the data of most immediate interest to food-process developers. It has been widely observed, however, that the interaction of solutes in the multicomponent mixtures usually increases the rejection or retention of solutes. Much more work will be necessary to determine what part of the observed selectivity is attributable to the membrane itself and what part to slow diffusion through a polarization boundary layer or a second "membrane" dynamically formed from rejected food constituents.

V. PHYSICAL PROPERTIES OF LIQUID FOODS

It has been pointed out that there is a well-established usage of concentrated foods of various kinds for both institutional and domestic consumption. To some extent this is bound to predetermine the nature of products made by reverse osmosis. The products will be seen by the consumer as replacements for or alternatives to already familiar forms of concentrated or dehydrated foods. Some familiar concentrated foods are tomato paste (30% solids), orange juice (40–50% solids), and milk (24–36% solids), which are all made by evaporative processes. It seems likely that processors will expect reverse-osmosis equipment to be capable of achieving a comparable degree of concentration. Whether this is an easy or a difficult problem depends upon three physical properties of the liquid, namely, its osmotic pressure, its viscosity, and the mass diffusivity of the solutes in the liquid.

A. Osmotic Pressure

In single-stage reverse osmosis it is necessary that the hydrostatic pressure in the apparatus exceed the osmotic pressure of the material being concentrated. Relatively few precise data on the osmotic pressure of foods are available. Also, because of the variability of natural materials, accurate measurements for one sample could be used only as an approximation for another sample of the same commodity. (A notable exception is the class of refined sugars, one of the few cases for which reliable data are available.) Estimates of osmotic pressure can be obtained by use of a well-known

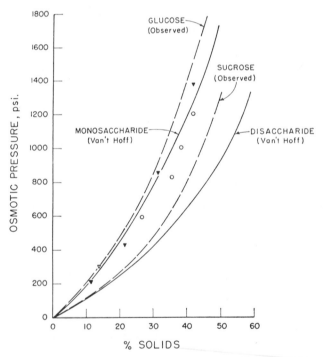

Figure 5.   Osmotic pressures of fruit juices and sugar solutions at 20–25°C. Experimental data for glucose,[17] sucrose,[13, 45] apple juice (▼)[34] and orange juice (○, ▼).[11, 34]

thermodynamic approximation, the van't Hoff equation, in the case of solutions whose components are of known molecular weight. The relation between osmotic pressure and composition is shown in Figure 5, for both the ideal dilute solution (van't Hoff equation) and real behavior of sugars in water. With respect to osmotic pressure, "sugary" foods (such as fruit juices, purees, and milk) behave as if they were composed of solutes which have an average molecular weight between 180 and 360. "Proteinaceous" materials (e.g., egg) have much lower osmotic pressures.

B. Diffusivity of Water and Solutes in Liquid Phase

Separation of water and solutes occurs near the surface of the reverse-osmosis membrane; water permeates into the membrane and solutes are rejected. The separated solutes would accumulate indefinitely at the membrane surface and the separation process would grind to a halt if some mechanism were not provided for the continuous removal of the solutes. Solute removal (or depolarization) can occur by three mechanisms: bulk flow, turbulent mixing, and molecular diffusion. It is known (see for example, Bird[4] or Schlichting[47]) that even when there is considerable eddy motion in the center of a flow channel, near the wall there exists a boundary layer in which turbulence is suppressed and mass transport occurs mainly by molecular diffusion and orderly (streamline) bulk flow. Furthermore, bulk velocities in this region are relatively small, and therefore a significant amount of the total transport occurs by molecular diffusion. The rate of diffusional transport is expressed in terms of a diffusion coefficient, or diffusivity. Diffusion coefficients of solutes and water in liquid foods are of the order of 10^{-7}–10^{-5} cm^2/sec, tending to be lower at high solute concentrations and high viscosity. Unfortunately, very little exact data are available for any liquid food material except sugar syrups. (A few data are given in Table 3.) A semi-empirical expression (Equation 1) relating diffusivity, concentration, and viscosity in liquids was derived by Gordon.[15]

$$D = D_0 \left(\frac{\mu_0}{\mu}\right) \left[1 + \frac{d \ln \gamma}{d \ln c}\right] \quad (1)$$

where D = mass diffusivity, cm^2/sec
D_0 = diffusivity at infinite dilution
μ = viscosity of the solution
μ_0 = viscosity of the pure solvent
γ = activity coefficient of solute
c = concentration of solute

Gordon's equation indicates that any material that demonstrates a large

Table 3.   Solute–Water   Diffusion   Coefficients   for   Some   Liquid Foods   at   25°C

Solute	Solute Concentration	Diffusion Coefficient (cm^2/sec)	Reference
Orange juice solids	42° Brix[a]	0.03×10^{-5}	11
	18° Brix	0.05×10^{-5}	
	0° Brix[b]	0.13×10^{-5}	
Sucrose	65%	0.083×10^{-5}	18
	42%	0.19×10^{-5}	
	18%	0.35×10^{-5}	
	0%	0.52×10^{-5}	
Ovalbumin	0.8%	7.6×10^{-7}	43

[a] Solids diffusing from solution of this concentration into water.
[b] Extrapolated to infinite dilution.

increase in viscosity with concentration (as most foods do) will suffer a corresponding decrease in diffusivity. Any such material will thus tend to exhibit severe concentration polarization and correspondingly low flux.

C.   Viscosity

Users of tomato catsup and maple syrup are aware that most foods do show a marked increase in viscosity with increased solids content. Since it is necessary to pump the material to be concentrated into and through reverse-osmosis apparatus, often through quite narrow apertures, some appreciation of the difficulties due to viscosity changes is essential.

The viscous properties of a typical sample of tomato puree are shown in Figure 6, where it is seen that the apparent viscosity is a function of the rate of shear as well as of the concentration. (These data should be taken as illustrative only, since measurements on other samples, or even repeat determinations on the same sample would probably give somewhat different results.) Thus, in the language of rheology, many foods are non-Newtonian fluids. Those that contain suspended solids are likely to be pseudoplastic and time-dependent fluids. Precise rheological data are available for sugar syrups, but with few exceptions[24, 46] data are scarce for other foods.

The effect of the increase of viscosity with concentration is occasionally to limit the practical degree of concentration that can be attained, where this is not otherwise limited by osmotic pressure. Furthermore, the increase in viscosity acts generally to aggravate the phenomenon of concentration polarization (see below) and thereby becomes in many instances the factor that ultimately determines what the permeation rate will be.

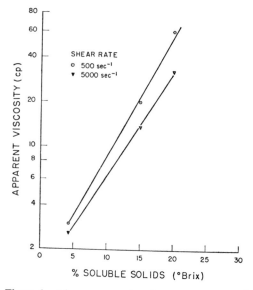

% SOLUBLE SOLIDS (°Brix)

Figure 6. Viscous properties of tomato puree at 25°C. The apparent viscosity is a function of both concentration and conditions of shear. Data obtained with tube viscometer;[31] commercial canned tomato juice concentrated by reverse osmosis.

VI. LIQUID PERMEATION RATE

A detailed discussion of the general factors that influence water permeation rate may be found in Chapter 8. At this point it will be sufficient to point out why permeation rates obtained with foods may be considerably lower than those obtained in other applications of reverse osmosis, such as desalination, and to outline the special techniques required for calculating these rates.

Equation 2 gives, in somewhat simplified form, the relation between the flux through a semipermeable membrane, the pressure in the system, and the concentration difference across the membrane.

$$J = \frac{1}{R_m}(\Delta P - \Delta \pi) \tag{2}$$

where J = water flux through membrane
ΔP = hydrostatic pressure difference across the membrane
$\Delta \pi$ = osmotic pressure difference across the membrane
R_m = resistance of the membrane to water permeation

Because of the rejection of solute at the membrane surface, the solute concentration there, and hence the local osmotic pressure, tends to be higher than in the bulk of the solution. The result is a decrease in water flux. This buildup of solute near the membrane (concentration polarization) is counteracted by diffusion of solutes into the bulk of the fluid, by turbulent eddies, and by shearing of the fluid near the membrane. *For foods, the increase in viscosity, mentioned earlier, interferes with all these mechanisms and thus leads to aggravated polarization and much reduced flux.*

Strictly speaking, polarization reduces the driving force in Equation 2 by increasing π_w, the osmotic pressure at the membrane surface. For illustrative purposes we can still consider the original overall driving force, that is

$$\Delta P - \Delta \pi = \Delta P - (\pi_{\text{bulk}} - \pi_{\text{permeate}})$$

and consider that polarization results in an additional resistance to flux, originating in the fluid and acting in series with the resistance of the membrane. Then

$$J = \frac{1}{R_m + R_f} (\Delta P - \Delta \pi) \tag{3}$$

where R_f is the additional polarization resistance.

Several authors[3a, 3b, 9, 12, 14, 33] have attempted to evaluate the polarization resistance both by experiments and by mathematical modeling of reverse-osmosis systems, and have concluded that in many instances involving viscous food materials, the *polarization resistance R_f is dominant in determining the permeation rate.* This is true even at rather low fluxes, so that a membrane capable of passing, say, 20 gal/(ft²)(day) of water at 1000 psi, may pass 5 gal or less of permeate from a dilute fruit juice or a sugar solution. The rate may dwindle to a value as low as 1 gal/(ft²)(day) if the concentrate approaches the consistency of tomato paste.

The magnitude of the polarization resistance is dependent upon the properties of the fluid, the condition of the flow, and the geometry of the apparatus. The interrelationship of these variables is quite complex and even estimates of the permeation rate are difficult and tedious to obtain. An approximate procedure has been worked out, however, based on a number of numerical solutions to transport equations. In the next section we briefly review the mathematical development of the procedure that can be used to calculate the expected permeation rates for practical problems.

A. Water and Solute Transport in the Boundary Layer

Consider a liquid food being concentrated in a reverse-osmosis channel formed by two parallel membranes (shown in Figure 7). If it is assumed that a solute is transported in the channel only by orderly (laminar) convection

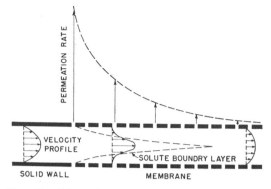

Figure 7. Schematic cross section of a long reverse-osmosis channel. The buildup of solute in the boundary layer distorts the velocity profile and reduces the permeation flux.

or lateral diffusion, then a differential mass balance will produce Equation 4:

$$\frac{\partial(uc)}{\partial x} + \frac{\partial(vc)}{\partial y} = \frac{\partial}{\partial y}\left(D\,\frac{\partial c}{\partial y}\right) \tag{4}$$

where c = solute concentration

x = coordinate along the axis of the channel
y = coordinate normal to the membrane
u = fluid velocity in x direction
v = fluid velocity in y direction
D = diffusion coefficient

At constant density, Equation 4 reduces to 5:

$$u\,\frac{\partial c}{\partial x} + v\,\frac{\partial c}{\partial y} = \frac{\partial}{\partial y}\left(D\,\frac{\partial c}{\partial y}\right) \tag{5}$$

It is convenient (and instructive) to express Equation 5 in dimensionless form, as in

$$u^+\frac{\partial c^+}{\partial x^+} + v^+\frac{\partial c^+}{\partial y^+} = \frac{\partial}{\partial y^+}\left(\frac{D}{D_w}\,\frac{\partial c^+}{\partial y^+}\right) \tag{6}$$

The dimensionless variables are defined as follows:

$$u^+ = \frac{u}{\bar{u}_0}$$

where $\bar{u}_0$ is the average bulk velocity at the channel entrance.

$$v^+ = \frac{vY}{2D_w}$$

where Y is the distance between the two parallel membranes which form the channel walls, and D_w is the magnitude of the diffusion coefficient at the surface of the membrane.

$$x^+ = \frac{4D_w x}{\bar{u}_0 Y^2}$$

$$y^+ = \frac{2y}{Y} - 1$$

$$c^+ = \frac{c_w - c}{c_w - c_0}$$

where c_w is the concentration at the membrane surface and c_0 is the average bulk concentration at the channel entrance. It can be shown that when solutions of Equation 6 are obtained by boundary-layer analysis, and the flux is calculated, the result will take the form

$$\frac{2\bar{J}Y}{D_w} = f\left(\frac{4D_w X}{\bar{u}_0 Y^2}, c_w, c_0\right) \qquad (7)$$

where $\bar{J}$ is the permeation rate averaged over the length of the channel and X is the length of the channel.

In the general case, the wall concentration, c_w, is not known. However, in the special case in which the membrane resistance to water transport is negligible compared to the resistance in the fluid in the channel, the solute concentration at the wall approaches the osmotic equilibrium concentration at the pressure of the system. (This situation occurs when there is no need for a large driving force for transport through the membrane itself, hence no difference in activity across the membrane.) In this situation, numerical solutions of Equation 6 can be obtained by various techniques, provided that the velocity distribution across the channel can be specified. The latter may be done in the manner described in the following section.

B.   Fluid Velocity Distribution Across the Channel

The viscosity of many food materials is such that their flow in thin unobstructed channels is characterized by Reynolds numbers of the order of 10^3 or less. Thus the flow should be laminar, and if the rate of permeation is small (or zero) the velocity distribution will be substantially parabolic, as suggested in Figure 7 at the entrance and the exit of the channel.

If, however, the rate of permeation is large enough to lead to significant increase in viscosity near the channel walls, the rate of shear will be reduced accordingly, and the velocity profiles will be distorted, as shown in the central velocity profile in Figure 7. An estimate of the actual velocity distribution can be obtained, if the rheological equation of state is known, and if a reasonable assumption about the distribution of shear stress can be made.

A number of rheological models have been proposed. One of the simpler ones is the so-called power law, Equation 8.

$$\tau = -a \left|\frac{du}{dy}\right|^{n-1} \frac{du}{dy} \tag{8}$$

where τ = shear stress (dynes/cm^2)
a = consistency index (empirical constant)
n = flow behavior index (empirical constant)

In the range of shear stress of interest, the power law describes with fair accuracy the behavior of a number of fluid foods, provided it is expressed in concentration-dependent form. This may be done by rearranging Equation 8 and expressing the parameters a and n as functions of composition. The result is Equation 9:

$$\frac{du}{dy} = \frac{\tau |\tau|^{\alpha\left(\frac{s}{1-s}\right)}}{\mu_s e^{\beta\left(\frac{s}{1-s}\right)}} \tag{9}$$

where s = weight fraction solids in the liquid food
μ_s = viscosity of pure solvent (water)
α, β = empirical constants

If, now, it is assumed that the magnitude of the shear stress is distributed linearly across the channel (an assumption that is exactly true in the case of zero flux) the axial velocity distribution can be calculated by integration of Equation 9, as indicated in 10.

$$u(y_1) = \int_0^{y_1} \frac{\tau |\tau|^{\alpha\left(\frac{s}{1-s}\right)}}{\mu_s e^{\beta\left(\frac{s}{1-s}\right)}} dy \tag{10}$$

where y_1 is any point across the channel.

The lateral velocity distribution can be estimated through the equation of continuity, Equation 11:

$$\frac{\partial u}{\partial x} + \frac{\partial v}{\partial y} = 0 \tag{11}$$

C. Magnitude of the Water Flux

Lengthy numerical solutions of Equation 6 indicate that the results take the approximate form of Equation 12 over much of the range of interest.

$$\frac{2\bar{J}Y}{D_w} = \phi \left[\frac{D_w x}{Y^2 \bar{u}_0}\right]^{-\frac{1}{3}} \tag{12}$$

The flux coefficient ϕ is a function of the rheological parameters α and β and of the solute concentration both at the inlet (c_0) and at the wall (c_w). Values of ϕ are reproduced in Table 4 for several concentrations and values of β. Intermediate values can be obtained by linear interpolation.

Table 4.   Flux Coefficient ϕ For Use in Equation 12

c_0	c_w	$\beta =$	0	2	4	6	8	10	15	30
0.05	0.10		2.35	2.32	2.28	2.24	2.21	2.17	2.10	1.90
	0.20		4.83	4.60	4.40	4.22	4.06	3.92	3.62	3.05
	0.30		6.34	5.86	5.46	5.14	4.87	4.64	4.21	3.48
	0.40		7.45	6.67	6.08	5.64	5.30	5.02	4.52	3.70
	0.50		8.34	7.22	6.48	5.95	5.56	5.25	4.70	3.82
0.10	0.20		2.35	2.27	2.20	2.13	2.06	2.00	1.87	1.60
	0.30		3.79	3.53	3.31	3.13	2.97	2.84	2.58	2.13
	0.40		4.83	4.35	3.98	3.70	3.47	3.30	2.96	2.43
	0.50		5.66	4.92	4.41	4.06	3.79	3.58	3.21	2.61
0.20	0.30		1.37	1.31	1.25	1.21	1.17	1.13	1.05	0.88
	0.40		2.35	2.16	2.01	1.88	1.78	1.69	1.52	1.25
	0.50		3.14	2.77	2.50	2.30	2.15	2.03	1.82	1.48
0.30	0.40		0.97	0.92	0.88	0.84	0.81	0.78	0.71	0.59
	0.50		1.73	1.56	1.43	1.33	1.25	1.18	1.06	0.86
0.40	0.50		0.75	0.71	0.67	0.63	0.61	0.58	0.53	0.43

The coefficients listed in Table 4 were obtained using an approximate velocity profile which ignores the non-Newtonian character of the fluid. That is, the integration in Equation 10 was performed by setting α equal to zero and treating the fluid as a concentration-dependent Newtonian fluid. This approximation reduces the computer time necessary for a solution by several orders of magnitude. Deviations from the more complete solutions are not serious and therefore the values of ϕ in Table 4 can be used for any value of α.

Table 4 and Equation 12 may be used for making preliminary design calculations. The technique is illustrated in the Appendix. Typical results,

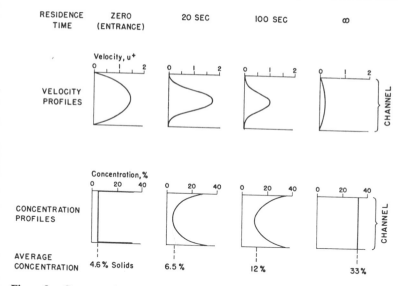

Figure 8. Concentration profiles and velocity profiles in transverse section of food liquid as it is concentrated during passage through long reverse-osmosis channel.

such as those in Figures 7 and 8, qualitatively illustrate why design of a reverse-osmosis plant for food applications is so complex compared to desalination.

Figure 8 shows in four stages what happens when tomato juice is concentrated at 1000 psi in a long channel. At zero residence time the juice enters the channel at 4.6% dissolved solids. Removal of water through the membrane surface immediately raises the concentration at the walls to 33% dissolved solids. (At 33%, tomato juice is assumed to have an osmotic pressure of 1000 psi, the operating pressure.) Since the concentration profile is still flat, except at the membranes, the velocity profile exhibits a normal laminar flow shape at the entrance.

Continual removal of water along the flow path and back diffusion of the solutes toward the center of the channel produce the concentration profile shown in Figure 8 after 20 sec of residence time. The wall concentration is still 33% and the center-line concentration is just beginning to rise above the original concentration. The velocity has slowed in the viscous region near the wall and has increased along the centerline of the channel where the juice is still dilute. The average concentration is 6.5%.

By the time the juice has traveled down the channel for 100 sec, the average concentration has risen to 12%. The juice is moving very slowly in the region near the membranes and the entire velocity profile is reduced because of loss of water. If the channel were infinitely long, the average concentration would eventually reach 33%, permeation would cease, the concentration profile would be flat, and the velocity profile would again have a shape characteristic of undisturbed laminar flow.

As was illustrated in Figure 7, the result of these phenomena is a permeation rate which rapidly diminishes along the channel length. Equipment designers have attempted to compensate for this in ways enumerated in the next section.

VII. APPARATUS

As a consequence of the peculiar properties of foods, namely, high osmotic pressure, high viscosity, lability, and susceptibility to microbiological attack, special apparatus is needed. The general requirements for food use are great mechanical rigidity to withstand pressures up to and exceeding 1000 psi, large membrane area because of low flux, some means to minimize concentration polarization without excessive pressure drop, and convenient disassembly for inspection and cleaning. In addition each food product presents unique difficulties. For example, a fruit juice may have large amounts of suspended particles, cheese whey has a tendency to form a sediment or gel on the membrane surface,[22, 23, 41, 42,] protein is damaged by excessive shear (as in pumps or discharge valves), and egg white has a normal pH of about 9. At the present time most available commercial units are compromises between meeting these requirements and achieving manufacturing simplicity and low equipment cost.

One configuration that has been widely used in food pilot plants is the round tube. This configuration is especially valuable if the product contains suspended matter and is not too viscous so that turbulent flow can be achieved. With laminar flow, however, severe concentration polarization occurs with the $\frac{1}{2}$–1 in. diameter tubes which have been tried. Attempts to improve the permeation rate by various turbulence promoters have met with considerable success. The devices used are simple and effective: discs perpendicular to the flow path spaced along the tube on a central rod; plastic balls filled into a tube of about the same diameter so that the balls undulate in the flow stream; and volume displacer rods which direct the flow toward the membrane and down the tube in a spiral path. Unfortunately with tubes, access to the interior for inspection and cleaning is a problem, particularly if a sediment is formed that requires physical removal. At present most manufacturers rely on cleaning in place; one recommends pumping a spongelike plug through the entire series of open tubes.

The spiral-wound module, which has been used for concentrating maple sap, has many of the features that lead to good performance in laminar flow. The "brine" side is essentially a short thin channel between parallel membranes. The brine-side spacer further increases "turbulence" or mixing in clear liquid foods. A major drawback, however, is that suspended particles clog the spacer and the modules are difficult to clean. Willits[21] has used an in-line germicidal ultraviolet lamp to minimize growth of microorganisms in the products being concentrated in these modules.

Other thin-channel devices and hollow-fiber units are attractive theoretically for concentrating viscous foods. In practice their use has been limited by unanswered questions concerning bacteriological growth and cleaning.

One design approach, specifically for food use, is that of Lowe et al.[28] This device (Figure 9) consists of a simple plate-and-frame type of assembly that is readily dismantled for access to each membrane. The fluid to be concentrated is directed acrosss a series of flat membranes spaced about 0.060 in. apart. Permeate seeps through the membranes, and then flows laterally through the mesh of stainless-steel wire cloth membrane supports, to be collected at the periphery of the device. As many as 20 membranes in series have been used in the stack assembly. Internal pressure is maintained by a pneumatically activated positive-displacement feed pump. The stack has been operated successfully at pressures up to 1500 psi. The fluid being processed can be recirculated within the high-pressure section of the system. After the desired concentration is attained, control of the concentration ratio is achieved by means of a proportioning system which withdraws product at a rate proportional to the rate of production of permeate. Another feature is that the product is discharged without shear damage.[27] Unfortunately this stack arrangement has not been manufactured commercially.

A. Improving Performance

Factors affecting equipment design and operation for food and other biological materials have been discussed in the literature.[5, 19, 32, 33] Here we will attempt only to summarize the important conclusions. Thinner channels in laminar flow result in markedly higher permeation rates because polarization is reduced. Increasing the intrinsic permeation rate of the membrane will have less effect, except for the most carefully designed equipment, because the major resistance to water transport is in the polarization layer, not the membrane. Higher operating pressures do not always increase flux but may only aggravate polarization, increase pumping costs, compact the membrane or a fouling layer of food solids, or even damage membranes and membrane supports. Recycling, that is, removing less water per pass through the apparatus but returning a portion of the concentrate to the feed stream,

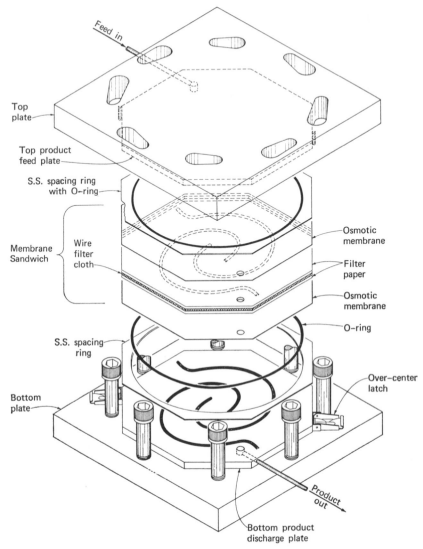

Feed in

Top plate

Top product feed plate

S.S. spacing ring with O-ring

Membrane Sandwich

Wire filter cloth

Osmotic membrane

Filter paper

Osmotic membrane

O-ring

S.S. spacing ring

Over-center latch

Bottom plate

Product out

Bottom product discharge plate

Figure 9.   Plate-and-frame reverse-osmosis device developed by USDA for food use.

212

sometimes improves the overall rate because the velocity profile in the channel is more uniform and relatively stagnant layers near the membrane are thinner.

Velocity is generally an important variable in reverse osmosis but care must be exercised in predicting the effect of velocity in food processing. Increased velocity helps prevent sedimentation and thus aids in achieving long-term operation (particularly if turbulent flow can be achieved). One reason that a given piece of equipment gives higher permeation rates at higher velocities is that the product does not reach as high a final concentration. This means that in order to achieve the required solids concentration, the flow path must be made longer at higher velocities. The permeation rate will be very low in the added length (because the feed is concentrated by then). The net result is that for laminar flow and a fixed amount of water removal, the average permeation rate is the same whether one uses a high velocity and a long flow path or a low velocity and a short flow path. On the other hand, high velocities do cause high pressure drop, hence loss of driving force and reduction in flux.

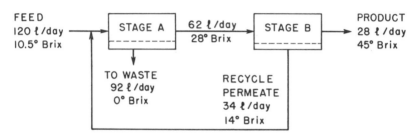

Figure 10.   Two-stage pilot-scale concentrator for orange juice.[10] With the concentrations shown, stage A would have to operate above 600 psi and stage B above 1100 psi.

Turbulence promoters or mixing devices may give improved performance, the increase in cost of the more sophisticated equipment being offset by improved flux and less membrane area. Some devices were mentioned above. In addition, Lowe[26] has experimented with pulsed flow and pulsing plastic spheres in attempts to increase permeation rates.

Finally, a two-stage system may be desirable to reduce the pressures necessary to achieve high solute concentration.[10, 19, 32, 33] An example is shown in Figure 10. In one stage part of the solutes are allowed to pass through a relatively nonselective membrane to reduce the osmotic pressure difference across the membrane and thus reduce the hydrostatic pressure needed to produce reasonable flow rates. Solutes in the permeate from this stage must be recovered in another stage operating at a lower concentration.

APPENDIX. DESIGN CALCULATIONS

The following numerical example is given in order to illustrate the use of Table 4 and Equation 12 for estimating the size of equipment needed to concentrate a viscous fluid.

Problem

A fruit juice is to be concentrated from an initial solids content of 10% to a final concentration of 35%. The apparatus is to be operated at a pressure of 1000 psi, which is known to be the osmotic pressure equivalent to a concentration of 42% solids in this particular material. The apparatus is to be a plate-and-frame type, consisting of flat parallel membranes, arranged for series flow, as shown in Figure 9. The length of the fluid path (x) on each set of membranes is to be 150 cm. Depth of the flow channels (Y) is to be 0.2 cm. Fluid velocity at the inlet to the apparatus is to be 5 cm/sec.

The pertinent physical properties of the fluid are

viscosity	$\mu = 0.01 \exp 8(s/1 - s)$(poise) (i.e., $\beta = 8$)	(1a)
specific gravity	$\rho = 1/(1 - 0.37s)$(g/cm³)	(2a)
diffusion coefficient	$D_w = 0.5 \times 10^{-5}$ (cm²/sec)	(3a)

where s is the weight fraction of solids in the solution (e.g., 0.10 for the feed). Answers to the following question will be obtained:

1. What will be the overall permeation rate, gal/(ft²)(day)?
2. How many steps (membrane pairs) are required to reach the desired concentration of 35% solids (0.40 g/cm³)?
3. What is the pressure drop through the apparatus?

Calculation Procedure

The first step in the procedure is to compute the fluid concentration at the membrane surface (42% solids). Equation 2a is used to convert this to the units of Table 4 (i.e., g/cm³).

$$\rho = 1/(1 - 0.37 \times 0.42) = 1.184$$
$$c_w = s \times \rho = 0.42 \times 1.184 = 0.50 \text{ g/cm}^3$$

The next step is to prepare a table such as Table 5, which shows the amount of concentration which would be achieved *per step* at various levels of solute

Table 5.   Concentration Change Per Step in Sample Problem

Step Inlet Concentration, c_0' (g/cm^3)	ϕ (Table 4)	Step Inlet Velocity, $\bar{u}_0'$ (cm/sec) (Equation 4a)	$\dfrac{D_w x}{Y^2 \bar{u}_0'} \times 10^3$	$\dfrac{2\bar{J}Y}{D_w}$ (Equation 11)	Δc (Equation 6a)
0.10	3.79	5.0	3.75	24.4	0.0101
0.15	2.88	3.3	5.63	16.2	0.0151
0.20	2.15	2.5	7.50	11.0	0.0180
0.25	1.63	2.0	9.38	7.7	0.0194
0.30	1.25	1.7	11.3	5.6	0.0200
0.35	0.87	1.4	13.1	3.7	0.0168
0.40	0.61	1.3	15.0	2.5	0.0106
0.50	0			0	0

concentration. In order to perform the calculation it is necessary to refer to Table 4 to obtain the constants for Equation 12, and also to make use of Equations 4a and 5a below. (Equations 4a and 5a are based on material balances for the solute and the solution as a whole. The derivations are straightforward but the latter one is fairly lengthy, and therefore will not be given here.)

$$\frac{\bar{u}_0'}{\bar{u}_0} = \frac{c_0}{c_0'} \tag{4a}$$

$$\Delta c = c_0' \left[\frac{\rho_x}{\rho_0 - \left(\dfrac{2\bar{J}Y}{D_w}\right)\left(\dfrac{D_w X}{\bar{u}_0' Y^2}\right)\rho_s} - 1 \right] \tag{5a}$$

where $c_0' =$ any concentration at inlet to step
$\bar{u}_0' =$ velocity at inlet to step
$\rho_x =$ density at end of step
$\rho_0 =$ density at inlet to step
$\rho_s =$ density of permeate

Because of the linear relationship between ρ and c (Equation 2a), Equation 5a reduces to the convenient form

$$\Delta c = c_0' \left[\frac{1}{1 - \left(\dfrac{2\bar{J}Y}{D}\right)\left(\dfrac{Dx}{\bar{u}_0' Y^2}\right)\rho_s} - 1 \right] \tag{6a}$$

The concentration changes (column 6 of Table 5) are then plotted against the inlet concentrations, as shown in Figure 1a.

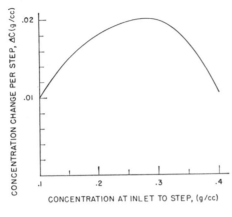

Figure 1a. Concentration change expected for a single pair of membranes at various concentration levels under conditions of sample problem.

Table 6. Step Calculations for Sample Problem

Step	c_0' (g/cm^3)	Δc	$\dfrac{s}{1-s}$	μ (cP)	$\bar{u}_0$ (cm/sec)	$\mu\bar{u}_0$
1	0.10	0.0100	0.1067	2.35	5.00	11.7
2	0.1100	0.0111	0.1182	2.57	4.54	11.7
3	0.1211	0.0122	0.1299	2.83	4.13	11.7
4	0.1333	0.0134	0.1455	3.20	3.75	12.0
5	0.1467	0.0145	0.1616	3.64	3.40	12.4
6	0.1612	0.0154	0.1794	4.20	3.10	13.0
7	0.1766	0.0166	0.1987	4.90	2.83	13.9
8	0.1932	0.0174	0.2200	5.81	2.59	15.0
9	0.2106	0.0184	0.2428	6.98	2.37	16.5
10	0.2290	0.0186	0.2676	8.50	2.18	18.5
11	0.2476	0.0196	0.2934	10.45	2.02	21.1
12	0.2672	0.0199	0.3213	13.07	1.87	24.4
13	0.2871	0.0200	0.3505	16.51	1.74	28.7
14	0.3071	0.0198	0.3808	21.04	1.63	34.3
15	0.3269	0.0192	0.4117	26.94	1.53	41.2
16	0.3461	0.0178	0.4426	34.49	1.44	49.7
17	0.3639	0.0160	0.4721	43.68	1.37	59.9
18	0.3799	0.0139	0.4994	54.33	1.32	71.7
19	0.3938	0.0120	0.5237	66.00	1.27	83.8
20	0.4058		0.5452			

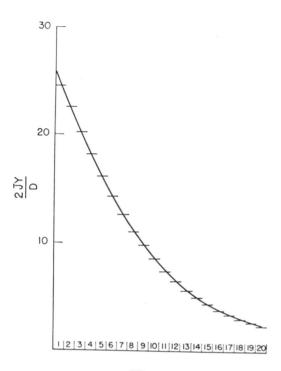

STAGE No.

Figure 2a.   Average dimensionless flux for each stage in sample problem.

Table 6 is constructed next. Columns 2 and 3 are obtained by use of Figure 1a, column 4 by reference to the specific gravity–concentration relationship (Equation 2a), and column 5 by use of the rheological equation of state (Equation 1a). Figure 2a is then constructed, using the dimensionless permeation rates from Table 5. These are plotted against the appropriate step numbers, which are obtained from Table 6, by inspection. The mean dimensionless permeation rate is obtained by summing the fluxes for each stage in Figure 2a and dividing by the number of stages. In this case nineteen steps are required to reach the desired outlet concentration and the overall mean permeation rate converted to real units is 2.7 gal/(ft²)(day).

The pressure drop through the apparatus is estimated by using the customary equation for pressure loss in laminar flow between flat, parallel surfaces, Equation 7a.

$$\Delta p = \frac{12\mu\bar{u}x}{Y^2} \tag{7a}$$

Equation 7a gives the pressure drop per step. The total pressure drop for the assembly (neglecting entrance and exit effects) is given by Equation 8a:

$$\Delta p_{\text{total}} = \frac{12N(\mu\bar{u})_{\text{avg}}X}{Y^2} \tag{8a}$$

where N is the number of steps. In this instance the total pressure drop was calculated to be about 4 psi, well within a tolerable range.

LIST OF SYMBOLS

a	Consistency index (equivalent to viscosity coefficient for Newtonian fluids)
c	Concentration of solute, g/cm^3
c_0	Concentration at entrance to channel
c_w	Concentration at membrane surface
c_0'	Concentration at entrance to a new step in a stepwise process
c^+	Dimensionless concentration, $(c_w - c)/(c_w - c_0)$
D	Mutual diffusion coefficient of solute in water solution
D_0	Diffusion coefficient of solute at infinite dilution
D_w	Diffusion coefficient at membrane surface
J	Permeation flux through membrane, $cm^3/(sec)(cm^2)$
$\bar{J}$	Flux averaged over length of channel
n	Flow behavior index
N	Number of steps in a stepwise process
ΔP	Hydrostatic pressure difference across the membrane
Δp	Pressure drop through reverse-osmosis channel
R_f	Fluid phase "resistance" to mass transfer (Equation 3)
R_m	Resistance of the membrane to water permeation
s	Weight fraction of soluble solids in liquid food
u	Fluid velocity in x direction
$\bar{u}$	Fluid velocity averaged across y
$\bar{u}_0$	Average fluid velocity at channel entrance
$\bar{u}_0'$	Average fluid velocity at entrance to a step in stepwise process
u^+	Dimensionless velocity, $u/\bar{u}_0$
v	Fluid velocity in y direction
v^+	Dimensionless lateral velocity, $vY/2D_w$
x	Coordinate along axis of channel
X	Length of channel
x^+	Dimensionless coordinate, $4D_w x/\bar{u}_0 Y^2$
y	Coordinate normal to the membrane
Y	Distance between membranes
y^+	Dimensionless coordinate, $(2y - 1)/Y$

α	Empirical rheological constant in Equation 9
β	Empirical rheological constant in Equation 9
γ	Activity coefficient of solute
μ	Viscosity of fluid
μ_s	Viscosity of pure solvent
π	Osmotic pressure
π_{bulk}	Osmotic pressure of fluid remote from membrane
$\pi_{permeate}$	Osmotic pressure of permeate
π_w	Osmotic pressure at membrane surface
$\Delta\pi$	Osmotic pressure difference across the membrane
ρ	Fluid density
ρ_0	Fluid density at inlet
ρ_s	Fluid density of permeate
ρ_x	Fluid density at any x
τ	Shear stress, dynes/cm^2
ϕ	Flux coefficient defined by Equation 12

REFERENCES

1. Abcor, Inc., Cambridge, Mass., *Bull. MS-68-8*, 1968.
2. Amicon Corporation, Cambridge, Mass., *Appl. Bull.* **102** (April, 1967).
3. Battelle Memorial Institute, Columbus, Ohio, *The Conference on Membranes for Industrial, Biological and Waste Treatment Processes, October 20–21, 1969*. (a) W. F. Blatt, A. Dravid, A. S. Michaels, and L. Nelson, "Solute Polarization and Cake Formation in Memhrane Ultrafiltration: Causes, Consequences, and Control Techniques." (b) R. P. de Filippi and R. L. Goldsmith, "Application and Theory of Membrane Processes for Biological and Other Macromolecular Solutions."
4. R. B. Bird, W. E. Stewart, and E. N. Lightfoot, *Transport Phenomena*, Wiley, New York, 1960.
5. W. F. Blatt, M. P. Feinberg, H. B. Hoppenberg, and C. A. Saravis, *Science* **150** (3693), 224–226 (1965).
6. W. F. Blatt, S. M. Robinson, F. M. Robbins, and C. A. Saravis, *Anal. Biochem.* **18** (1), 81–87 (1967).
7. *Canner Packer* **137** (10) (1968).
8. E. S. K. Chian and J. T. Selldorff, *Proc. Biochem.* **4**, 47 (Sept., 1969).
9. V. N. Das, "Reverse Osmosis: Sugar Solutions in Turbulent Flow Through Tubular Membranes," M.S. Thesis, University of California, Davis (1970).
10. C. Elata, "Development of Concentrator of Orange Juice by Reverse Osmosis," *Prog. Rep. 352-1*, Hydronautics-Israel, Ltd., Jan. 16, 1969, p. 14.
11. L. F. Ginnette, unpublished data.
12. L. F. Ginnette and R. L. Merson, "Maximum Permeation Rates in Reverse Osmosis Concentration of Viscous Materials," Paper No. 299, *National A.I.Ch.E. Meeting, St. Louis, Mo., February 18–21, 1968*.

13. S. Glasstone, *Textbook of Physical Chemistry*, 2nd Ed., van Nostrand, Princeton, N.J., 1946, p. 699.

14. R. L. Goldsmith, "Macromolecular Ultrafiltration with Microporous Membranes," paper presented at the *158th American Chemical Society Meeting, New York, September 7–12, 1969.*

15. A. R. Gordon, *J. Chem. Phys.* **5**, 522 (1937).

16. B. S. Horton, R. L. Goldsmith, S. Hossain, and R. R. Zall, "Membrane Separation Processes for the Abatement of Pollution from Cottage Cheese Whey," paper presented at the *Cottage Cheese and Cultured Milk Products Symposium, University of Maryland, College Park, March 11, 1970.*

17. *International Critical Tables*, Vol. 4, 1928, p. 430.

18. R. R. Irani and A. W. Adamson, *J. Phys. Chem.* **62**, 1517 (1958).

19. K. D. B. Johnson, J. R. Grover, and D. Pepper, *Desalination* **2**, 40 (1967).

20. S. Kimura, and S. Sourinajan, *Ind. Eng. Chem. Proc. Des. Dev.* **7**, 548–554 (1968).

21. J. C. Kissinger and C. O. Willits, *Food Technol.* **24**, 481 (1970).

22. T. H. Lim, "Role of Protein in Reverse Osmosis of Cottage Cheese Whey," M.S. Thesis, University of California, Davis (March, 1970).

23. T. H. Lim, W. L. Dunkley, and R. L. Merson, "Role of Protein in the Reverse Osmosis of Cottage Cheese Whey," *J. Dairy Sci.* **54**, 306–311 (1971).

24. E. S. Loureiro, M.S. Thesis, University of California, Davis (1970).

25. H. K. Lonsdale, U. Merten, and M. Tagami, *J. Appl. Polymer Sci.* **11**, 1807 (1967).

26. E. Lowe and E. L. Durkee, "Dynamic Turbulence Promotion in Reverse Osmosis Processing of Liquid Foods," *J. Food Sci.* **36**, 31–32 (1971).

27. E. Lowe, E. L. Durkee, R. L. Merson, K. Ijichi, and S. L. Cimino, *Food Technol.* **23**, 75 (1969).

28. E. Lowe, E. L. Durkee, and A. I. Morgan, Jr., *Food Technol.* **22**, 915 (1968).

29. P. G. Marshall, W. L. Dunkley, and E. Lowe, *Food Technol.* **22**, 969, 1968.

30. F. E. McDonough and W. A. Mattingly, *Food Technol.* **24** (1970).

31. R. L. Merson, unpublished data.

32. R. L. Merson and L. F. Ginnette, *Appl. Polymer Symp.* **13**, 309 (1970).

33. R. L. Merson, L. F. Ginnette, and A. I. Morgan, Jr., *Dechema Monograph.* **63**, 179 (1969).

34. R. L. Merson and A. I. Morgan, Jr., *Food Technol.* **22** (5), 631 (1968).

35. U. Merten, I. Nusbaum, and R. Miele, "Organic Removal by Reverse Osmosis," Gulf General Atomic, San Diego, California, *Rep. No. GA8744*, presented at *156th American Chemical Society Meeting, September 8–13, 1968, Atlantic City, N.J.*

36. A. S. Michaels, "Membranes: The Thin Difference," *Ind. Res.* (April, 1968).

37. A. S. Michaels, *Chem. Eng. Prog.* **64** (12), 31–43 (1968).

38. A. S. Michaels, "Ultrafiltration," in *Progress in Separation and Purification*, E. S. Perry, Ed., Vol. 1, Wiley-Interscience, New York, 1968, pp. 297–334.

39. A. I. Morgan, Jr., E. Lowe, R. L. Merson, and E. L. Durkee, *Food Technol.* **19**, 1790 (1965).

40. S. C. Palmer, "Pilot Plant Study for Modeling and Optimization of Reverse Osmosis with Cottage Cheese Whey," M.S. Thesis, University of California, Davis (July, 1970).

41. C. Peri and W. L. Dunkley, "Reverse Osmosis of Cottage Cheese Whey. 1. Influence of Composition of the Feed," *J. Food Sci.* **36**, 25–30 (1971).
42. C. Peri and W. L. Dunkley, "Reverse Osmosis of Cottage Cheese Whey. 2. Influence of Flow Conditions," *J. Food Sci.* **36**, 395–396 (1971).
43. A. Polson and G. M. Potgieter, *Nature* **204**, 379 (1964).
44. S. Sourirajan, "Characteristics of Porous Cellulose Acetate Membranes for the Separation of Some Organic Substances in Aqueous Solution," private communication (1964).
45. S. Sourirajan, *Ind. Eng. Chem. Proc. Des. Dev.* **6**, 154 (1967).
46. G. D. Saravacos, *Food Technol.* **22**, 1585 (1968).
47. H. Schlichting, *Boundary Layer Theory*, 4th Ed., McGraw-Hill, New York, 1960.
48. J. C. Underwood and C. O. Willits, *Food Technol.* **23**, 787–790 (1969).
49. *U.S. Dept. Agr. ARS 74-51* (Aug., 1969). Report of a symposium, "Reverse Osmosis in Food Processing," January 23, 1969. (*a*) W. L. Dunkley, "Concentrating and Fractionating Whey," pp. 19–28. (*b*) W. A. Mercer, "Treatment of Waste Streams from Food Processing," pp. 29–32. (*c*) R. L. Merson, "An Overview," pp. 1–10. (*d*) J. C. Underwood, "Concentration of Maple Sap," pp. 16–18.
50. *U.S. Dept. Agriculture, Agricultural Statistics*, 1968.
51. C. O. Willits, J. C. Underwood, and U. Merten, *Food Technol.* **21**, 24–26 (1967).
52. L. T. Zanto, L. M. Christopher, and S. E. Bichsel, "A Study of Some Physical and Chemical Parameters Affecting Non-sugar-sugar Partition with Cellulose Acetate Membranes," paper presented at *American Society of Sugar Beet Technologists, Phoenix, February 22, 1968.*

Chapter XI Applications for Reverse Osmosis in the Pulp and Paper Industry

Averill J. Wiley,* A. C. F. Ammerlaan,† and G. A. Dubey‡

1. INTRODUCTION

The development of feasible and practical applications of the technology of membrane processing to the water-quality problems of the pulp and paper industry has been the subject of extensive exploratory research and engineering studies in the period 1965–1970.§ Data from laboratory, pilot, and semi-commercial evaluations of reverse osmosis point to important large-scale applications for concentrating dilute wastes from pulp- and paper-making operations and for cleansing waters from individual unit operations so they can be recycled within the pulp mill or paper mill.[1-5] The high levels of rejection and of concentration of dissolved materials consistently demonstrated in these reverse-osmosis studies on pulp and paper waste waters provide a firm base for engineering design of new routes for effective control of water pollution.

It is important to emphasize the concept of "in-plant" treatment at this stage of developing pulp and paper industry applications for these membrane processes. Economic evaluations have not as yet favored the use of reverse osmosis as an "out-plant" method (i.e., processing the huge (millions of

* The Institute of Paper Chemistry, Appleton, Wisconsin.
† Abcor, Inc., Cambridge, Massachusetts.
‡ The Institute of Paper Chemistry, Appleton, Wisconsin.
§ Project initiated by Pulp Manufacturers Research League which merged into The Institute of Paper Chemistry effective April 1, 1970.

gallons per day) and extremely dilute flows obtained by combining and mixing the effluents from the many different and individual unit operations of a total mill complex).

Actually, the in-plant concept is not expected to be a limitation or a restriction upon the size and number of membrane process applications likely to be developed. This is true because the ultimate objective in plant design and operations for the pulp and paper industry and for other water-using industries must necessarily be directed to achieving a much greater degree of closing of the process-water system than is presently practiced. Counter-current washing can be an important first step to that end where applicable. Reduction of waste discharges by means of in-plant cleaning and reuse of process waters can be another important route to obtaining effective and complete answers to the problems of reducing the outflow of waste waters. Reverse osmosis is being developed as an effective new tool to be added to the kit of proven unit operations available to the process engineer to accomplish this objective of closing the mill water system.

II. SCOPE OF WATER TREATMENT AND WATER REUSE PROBLEMS OF THE PULP AND PAPER INDUSTRY

A trade directory[6] lists 528 pulp mills and 947 paper mills operating in the United States and Canada in 1969. Applications for reverse osmosis are most likely for processing the effluents from the washing and bleaching of pulps produced by chemical processes. The same source lists 198 chemical pulp mills including 116 alkaline sulfate (kraft) mills, 44 acid and high-yield sulfite mills, and 38 neutral sulfite semichemical (NSSC) mills in the United States. Possibilities for application of reverse osmosis have not yet been studied in detail for the less critical wastes of physical and physicochemical types of pulping in the United States which include 78 ground-wood, 10 chemical-mechanical, 32 roofing pulp, and 48 miscellaneous processes. Almost all chemical pulp mills designed and constructed in the past five years and many of the mills designed and built or rebuilt since 1955 are equipped with continuous or staged countercurrent washing and filter systems to concentrate pulp washing and bleaching wastes. On the other hand, the majority of older mills face multimillion dollar capital investments to concentrate and dispose of dissolved materials contained in their dilute effluents. Applications for reverse osmosis may be expected to develop eventually for effluents of various other unit operations than pulp washing in both newly designed and in older mills, but the need for concentration of pulp wash waters is especially critical in older plants, which must meet new standards of effluent quality before discharging the effluents into receiving waters. Mills in that category (perhaps 80% of the total number of operating plants, and producing upward of 50% of the total industry tonnage) are most likely to find one or several individual reverse-osmosis installations to be effective and economically justifiable alternatives for achieving the all-important reduction in volumes of wasted waters and for maintaining desired high standards of effluent quality for the remaining waste water outflows.

The Process and Disposal Committee of the Technical Association of the Pulp and Paper Industry compiled an authoritative listing[7] for the years 1965–1966 of the volumes of water intake and the amount of water reuse practiced in ten representative categories of the chief types of pulp- and paper-making operations. These data, presented in Table 1, show an intake of fresh process water as low as 0.4 kgal/ton in a paperboard mill using superior methods, and control of water usage to as high as 41 kgal/ton in a large 1100 tons/day kraft mill. The industry was credited with reusing its water intake about 2.5 times (ratio of reuse/intake in Table 1) in 1960 but has increased this average to more than three as reported in the latest available compilation. The research and development for applying reverse-osmosis processing is aimed toward making possible substantial increases in water reuse through recovery of 90% or more of effluent flows from individual unit operations as clean, clear, and reusable process water with concentration of the dissolved material to small manageable volumes more easily and economically processed for final disposal. Such processing can also provide for recovery and utilization of any materials of value originally contained in the waste water. This can be an important route to cost reduction for treating some of these dilute, wasted process effluents.

Pulp wash waters and bleach plant effluents are currently of critical concern to most chemical pulp mills. These streams are derived from a variety of substantially different pulping and bleaching systems, and may account for as much as 25% of the total load of organic matter coming from the pulp mill. These are alkaline kraft, acid sulfite, neutral sulfite, and high-yield bisulfite pulp wash waters. There are a variety of bleach effluents from various hypochlorite, chlorine, chlorine dioxide, and peroxide systems, and various acid and alkali extract stages of bleaching. In addition, there are barking waters, evaporator condensates, and various types and kinds of machine "white waters." When we then include the variety of hardwoods, softwoods, bagasse, and other raw materials, and the various degrees and levels of pulping and bleaching to produce the wide variety of types of pulp, paper, rayon, cellulose films, and converted products, we can readily list some 20–30 significantly different types of waste waters from different unit operations that are likely candidates for membrane processing in this industry.

Table 1. Water Intake and Reuse for Median and Superior Mills

Category	Supply Source (%) Ground	Surface	City	Production (tons/day)	Water Intake (10⁶ × gal/day)	Water Reused Total (10⁶ × gal/day)	Non-process	Breakdown by Process (%) Pulping	Bleaching	Stock prep.	Machine	Other Processes	Water Use (1000 gal/ton)	Intake (1000 gal/ton)
Bleached kraft														
Median	28	72	—	864	37	118	18	12	51	6	13	—	179	43
Superior	—	100	—	1100	45	229.5	13	8	53	9	17	—	249	41
Unbleached kraft														
Median	27	73	—	818	23.7	62.3	18	20	—	7	55	—	105	29
Superior	—	100	—	865	16.5	78.5	24.6	20.4	—	—	55	—	109.8	19.1
Sulfite														
Median	—	100	—	304	21	48.6	9	25	47	8	11	—	229	69
Superior[a]	—	—	—	130	5	24.5	1	35	13	4	47	—	227	38.5
Board, nonintegr.														
Median	19	77	4	215	3.35	7.1	16	—	—	26	55	3	48.5	15.5
Superior	—	100	—	300	0.12	5.3	6	—	—	33	28	33	18	0.4
Deinking														
Median	—	100	—	117	3.8	7.4	16	18	45	5	16	—	96	33
Superior	1	99	—	200	4.8	17.3	20	12	52	6	10	—	110.5	24
Soda pulp														
Median	11	89	—	300	21	39.8	15	5	22	—	58	—	203	70
Superior	—	100	—	350	23	103.3	3	5	24	—	68	—	361	66
Mechanical pulp														
Median	3	97	—	328	15.6	22.9	10	22	15	4	49	—	116	48
Superior	—	100	—	270	1.1	8	—	45	—	—	55	—	34	4
Semichemical pulp														
Median	20	80	—	363	9.8	13.2	18	14	23	3	42	—	63.5	27
Superior	—	100	—	177	1.73	7.97	3	25	—	11	61	—	54	9.8
Paper, nonintegr.														
Median	24.5	74.1	1.4	112	4	4.7	4	—	—	3	90	3	77.6	35.6
Superior	25	75	—	180	1.7	22.9	2	—	—	—	98	—	137	9.4
Misc. pulp														
Median	28	71	1	64	6.7	5	13	29	7	23	28	—	183	05
Superior[b]	50	50	—	160	1	6.2	2	3	—	43	42	—	45	16

a ... natural surface, no data were given.

III. AREAS FOR ECONOMIC APPLICATION OF REVERSE OSMOSIS

To determine potential economic applications for the reverse-osmosis process in the pulp and paper industry, we should look for areas in which the cost of reverse osmosis will be offset by positive benefits. Three different areas in which credits may arise from the application of the reverse-osmosis process are as follows:

1. Recovery of organic and inorganic values.
2. Concentration of pollutants for more economic disposal processing.
3. Recovery of clean reusable process water.

A. Recovery of Organic and Inorganic Values

A number of waste effluents from pulp mills contain organic and inorganic substances that have appreciable value. Values such as those outlined in Table 2, have been recognized for the effluents resulting from the pulp-washing operations that take place after the wood-digesting process.

Table 2. Possible Values in Concentration of Marketable Solids

Feed Concentration (% Solids)	Assumed Net Value of Solids (¢/lb Solids)	Value of Solids in Recovered Solutions at 90% Retention of Solids in Feed (¢/1000 gal Feed)
0.2	1	15
	2	30
	5	75
0.5	1	37
	2	75
	5	185
1.0	1	75
	2	150
	5	375

Most kraft, NSSC, and sulfite mills have installed evaporators for concentrating the stronger portions (8–16% solids) of the cooking liquors, which can readily be recovered and feasibly processed at this range of dissolved solids content. The evaporated liquor products, with approximately 50% solids, can then be sold either as is, or as dry powder after subsequent spray drying. Alternatively, the concentrated liquor may be burned for recovery of inorganics in the ash and of heat from the combustion of the organics. The cost of removal of water by concentration in medium-sized, multiple-effect evaporators now in use by the industry is in the range of $2–6/1000 gal of water removed, which makes water recovery expensive by that method and limits the application of evaporators to processing of the digester-strength liquors and the first-stage wash waters such that the combined feed concentration to the evaporator is seldom less than 8–10% solids.

Reverse osmosis, on the other hand, operates without phase change and is able to remove water from much more dilute wash waters at a cost that has been estimated to range between $1 and $2/1000 gal of water recovered, even with use of the presently available reverse-osmosis equipment in the early stages of commercial development. It is expected that costs can eventually be reduced well below $1/1000 gal. With the reverse-osmosis process there is a possibility of economically recovering additional amounts of valuable material from the dilute fraction from the washing process to supplement the recovery of digester liquors drained or displaced from the freshly cooked cellulose pulp. A similar situation, but with much less potential for recovery of marketable values, is being demonstrated in continuing studies upon some of the other effluents from a pulp mill (e.g., bleach plant effluents, evaporator condensates, and de-inking wastes).

B. Concentration of Pollutants for Economic Disposal

With more stringent laws controlling water pollution being placed in effect nationwide, more and more pulp and paper mills find themselves in a situation requiring higher levels of pollution abatement, particularly for reduction of the biochemical oxygen demand (BOD) of the waste waters they discharge. Industrial wastes from processing of organic material are often found to contain substantially higher concentrations of BOD_5 than the 25 mg/l range characteristic of domestic sewage. The cost of reducing the BODy of industrial waste effluents containing several thousands mg/l of BOD_5 b$_5$ biological oxidation processes can be expensive, as illustrated in Table 3.

In addition to the removal of BOD there is a strong trend by the authorities responsible for pollution control to require that color components, refractory organics such as lignin, and phosphates, nitrates, or other salts be removed from wastes to a greater degree than is now accomplished. These components are difficult to remove effectively, if at all, by biooxidation. Additional types and degrees of advanced waste treatment are required. Reverse osmosis can remove and concentrate all these different components in a single step, and therefore attracts increasing interest as a promising new method for treating waste waters.

Table 3. Cost of 90% BOD₅ Removal by Bio-oxidation at an Assumed Cost of 4¢/lb BOD₅ Removed

Feed Concentration BOD (mg/l)	Cost of BOD Removal ($/1000 gal Feed)
1000	0.30
2000	0.60
3000	0.90
4000	1.20
5000	1.50

C. Recovery of Clean Reusable Process Water

The waters recovered by reverse-osmosis processing of pulp-mill effluents are characterized by low levels of color and dissolved materials and by zero levels of suspended solids. The capability for reuse of these clean waters offers a significant process credit. The value to be assigned to the recovered water must be properly evaluated. Pulp mills are normally located so as to have large volumes of low-cost, clean water available and it is apparent that most pulp and paper operations can continue to have fresh treated water available in the foreseeable future at costs that are usually substantially less than 10¢/1000 gal. This is a fraction of the cost of reverse-osmosis permeate waters that can be obtained from the various types of membrane process equipment presently available in their early stages of commercial development. However, the pulp and paper industry is now being faced with revolutionary upward trends in the cost of processing outgoing waste waters to meet the requirements of effluent quality projected for 1975. Pulp and paper engineers are taking a new look at comparing the combined cost of treating raw water input and the waste treatment of effluent outflows against the alternative costs of closed system, in-plant treatment, and recycle of clean, renovated water.

There are a number of waste effluents of critical concern to the pulp and paper industry for which it seems probable that the application of reverse osmosis will provide a combination of benefits, such as removing BOD and other pollutants, producing marketable concentrates of dissolved solids, or recovery of clean, reusable water. This can be illustrated on the basis of a representative example for pulp wash water containing 1% solids (assumed net value of 1¢/lb), and 3000 ppm BOD₅. Treatment by secondary methods of biological oxidation would cost approximately $0.90/kgal (Table 3) and there would be no values recovered to help pay the cost of treatment. The application of reverse osmosis for the treatment of this type of waste water would result in the recovery of solids worth approximately $0.75/kgal of feed (Table 2) plus 900 gal of reusable water with an assumed value of $0.06. This means that there could be a credit of at least $0·80/kgal of feed toward the meeting of costs of the reverse-osmosis process. Furthermore, it demonstrates that as long as the total cost of reverse-osmosis operations does not exceed $1.70/kgal of feed, it could be economically advantageous to choose this process as an effective alternative to conventional biological oxidation methods of out-plant waste treatment.

Unfortunately, this opportunity for recovery of significant values is limited to a relatively few of the many pulp and paper process effluents. The relatively high costs of reverse osmosis cannot yet be balanced or supported as a means for pollution processing of other large volume wastes of this industry.

D. Effluent Streams that Should Be Considered

The general conclusion of the previous section, in which potential commercial applications for the reverse-osmosis process are discussed, is that initial applications for reverse osmosis will first become economically feasible for either dilute effluents from which values can be recovered to help meet treatment costs, or for effluents with high concentrations of pollutants that are processed with difficulty by other available methods, or both. With this in mind, we can develop in more detail the general conditions for application of the reverse-osmosis process in treatment of pulp- and paper-mill waste streams.

The following effluents should be considered as promising applications for the reverse-osmosis process:

- Pulp wash waters.
- Bleach plant effluents.

We have previously cited the above process wastes as especially critical and likely to become subject for concentration processing by reverse osmosis. Other waste streams receiving reverse-osmosis application studies include the following:

- Evaporator condensates containing volatile solutes, such as acetic acid.
- De-inking wastes.
- Dyewastes from manufacture of colored paper.

The daily volumes of these several process waste waters within the average-sized (100–500 tons/day) mill are fairly large. Advantage is readily apparent in reducing these volumes whenever possible by optimization of in-plant recirculation of water in order to reduce the size and cost of a reverse-osmosis plant.

Pulp washing and pulp methods and equipment may vary substantially, but modern mills are adopting countercurrent flow in one form or another. The flow sheet in Figure 1 illustrates a typical vacuum drum washer with

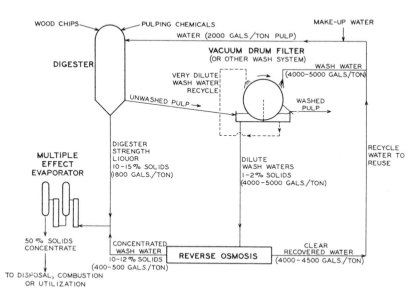

Figure 1.   Flow Sheet—pulping, pulp washing, wash water, and water reuse.

fresh water introduced in the final shower and dilute wash from that section recycled to the first shower. Two or more drum washers may operate in staged counterflow. New continuous digesters wash in the digester.

Evaporator condensates may contain volatile solutes of critical concern as water pollutants. In the case of evaporating spent liquors from acid sulfite pulping the condensates contain volatile organic acids, such as acetic and formic acids, and also sulfurous acid. Condensates from evaporation of alkaline (kraft) liquors contain methanol and also volatile sulfur compounds. These low-molecular-weight volatiles are not well rejected by most reverse-osmosis membranes now available. However, the salt forms of acetic and formic acid are quite well rejected and active application studies for reverse-osmosis treatment have reached pilot-scale study on neutralized condensates from acid pulping processes.

Preliminary evaluations indicate reverse osmosis may also have application to the processing of wastes from de-inking of printed waste paper. More study is indicated.

For each process water, a flow rate in the range of $0.2-2 \times 10^6$ gal/day or even more can be expected, with the concentration ranging from 0.1 to 1.5% total solids. At the present cost for reverse osmosis and for evaporation, the cross-over point from one method of concentration to the other is roughly at a point where the flux rate (J) of the reverse-osmosis process drops below 1-2 gal/(day)(ft²).

This low level of flux through the membranes results from the increased fouling and the rising osmotic pressure that accompany an increase in concentration of the dilute feed. Evaporation is not generally economical for pulp mill effluents with less than 8-10% solids; thus reverse osmosis should at least concentrate the dilute process waste waters up to that point, but preferably to even higher concentrations if the flux rate is satisfactory.

IV. GENERAL DESIGN CRITERIA

A. Concentrate of Dissolved Solids in Feed

As pointed out in the preceding paragraphs, the main design criterion for a commercial reverse-osmosis unit in the pulp and paper industry is to concentrate dilute effluents in the range of 0.5-1.5% solids to at least 8-10% solids and preferably higher. This means that reverse-osmosis systems must be designed for a concentrating factor of at least 5 and that the concentrating effect of the reverse-osmosis system may be as high as 20 or more times. Dilute effluents with less than 0.5% solids are common in the pulp industry but treatment of these very dilute effluents cannot be considered economically feasible at this early stage of developing the reverse-osmosis process. These very dilute effluents will probably have to be recycled or otherwise processed in the mill to the point that the concentration has increased to economically acceptable levels for concentrating by reverse osmosis.

B. Rejection Ratio

Once the desired concentrating ratio has been established, several other design factors can quickly be determined from Figures 2 and 3, which are based on the equations inserted in the figures. The lower part of Figure 2 is a graphical presentation of the relation between concentrating effect and amount of water to be removed from the initial feed. Membranes never exhibit 100% rejection for dissolved solids in a waste water, and the amount of water that must be removed to achieve a desired degree of concentration

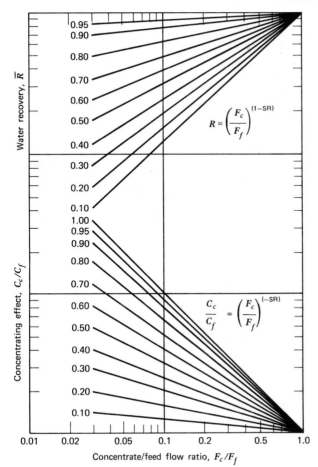

$$R = \left(\frac{F_c}{F_f}\right)^{(1-SR)}$$

$$\frac{C_c}{C_f} = \left(\frac{F_c}{F_f}\right)^{(-SR)}$$

Figure 2. Concentrating effect and water recovery vs. water removal for different rejection ratios (i.e., $1 - C_p/C$), represented by the numbers to the left of the lines.

depends on the extent to which solids are able to pass through the membranes. This, in turn, affects the area of membrane required. At the same time, the solids lost in the permeate result in less than 100% recovery of solids in the final concentrate stream of a reverse-osmosis system, which is important to the economics of a reverse-osmosis system. How rejection ratios and water removal affect the efficiency of a reverse-osmosis system as a recovery plant is shown in the upper part of Figure 2.

The average concentration and composition of permeate produced by a reverse-osmosis system designed for waste water treatment are also of prime

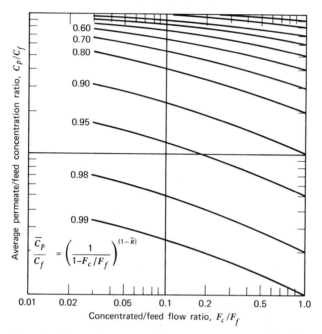

$$\frac{\overline{C}_p}{C_f} = \left(\frac{1}{1 - F_c/F_f}\right)^{(1-\overline{R})}$$

Figure 3. Permeate concentration vs. water removal for different rejection ratios (i.e., $1 - C_p/C$), represented by the numbers to the left of the lines.

importance. Figure 3 shows the correlations between water removal, rejection ratios, and concentration of permeate.

During several years of laboratory investigations certain generalized data were developed for rejection by reverse-osmosis membranes for pulp-mill effluents. These figures are presented in Table 4 and may serve as a guideline when using Figures 2 and 3 for a preliminary investigation of the feasibility of the reverse-osmosis process for the pulp industry.

Table 4.   Rejection Ratios[a] for Pulp-Mill Effluents

	Havens Type 3 Membrane (%)	Havens Type 5 Membrane (%)
Solids	75–98	95–99
Color	98–99.8	99–99.8
BOD	80–95	90–99
COD	85–98	90–99
Inorganics	70–95	94–99

[a] Rejection ratio $= 1 - c_p/c$, where C_p = concentration of solids in permeate and C = concentration of solids in concentrate stream.

C. Pretreatment and Selection of Equipment and Membranes

Once the design objectives for a reverse-osmosis plant have been established, the type of equipment design and the types of membrane must be selected before detailing of the design. In most cases, the approach is to select that combination of equipment design and membrane that will allow the reverse-osmosis process to operate with the least amount of pretreatment. In this respect, three aspects of pulp-mill effluents are important.

1. The presence of suspended solids and fibers in the dilute effluents.
2. The pH for these effluents, which ranges from 1.5 to 11.
3. The temperature of the effluents, which may vary from 10 to 85°C.

After much study, the 0.5-in.-diameter tubular membrane support has been selected from those commercially available at this stage of the technology of reverse-osmosis processing as the design best suited to handle effluents with suspended material. A simple step of removing large fibers and suspended particles by passing through a 100-mesh screen has proved to be satisfactory as the only pretreatment required in most pulp-mill applications. At present, all reverse-osmosis systems using tubular designs use cellulose acetate membranes. The rate of hydrolysis for cellulose acetate membranes is at a minimum at a pH of about 5. Deterioration of the cellulose acetate membranes increases as the pH ranges above or below this point. A lower pH limit of 3 and an upper limit of 7.5 seem to be economically acceptable in the pulp-mill applications so far studied. Effluents with a pH outside the range of 3–7.5 must be adjusted in pH before cellulose acetate membranes can be used. The hollow-fiber-bundle design, which uses nylon membranes, could be used for effluents with a pH higher than 7.5, if the effluents are carefully freed of any traces of suspended material. However, the advantage of the nylon membrane lies not only in reduced expense of pH effluents but also in preventing a substantial increase in membrane fouling that can result from precipitation of organics, which often occurs when the pH of alkaline or acidic pulp-mill effluents is adjusted over a considerable range.

For all reverse-osmosis systems currently on the market, whether based on cellulose acetate membranes or nylon membranes, there is a temperature limitation. The temperature of the effluent to be processed should not exceed an upper limit of 40–50°C. Hence, for certain pulp-mill effluents, installation of a heat exchanger ahead of a reverse-osmosis plant will be required for cooling. On the other hand, the temperature of the effluent should be as high as possible in order to obtain maximum flux rates. For a solution that does not cause membrane fouling, the flux rate (J) improves by about 2.8 % °C increase in temperature:

$$J = [(t - 25)(2.8)](J_{25})$$

where t = operating temperature, °C

J_{25} = flux at 25°C

Experience has shown that fouling by organic materials becomes an increasingly critical problem at lower temperatures. This is apparent when resinous woods, such as pine, are being pulped. Pitch, which may be in colloidal solution, may separate and deposit on the membranes if the temperature is too low, especially at the lower levels of velocity or turbulence within the reverse-osmosis system.

D. Membrane Area and Module Configuration as Related to Flux Rate and Minimum Velocities

After determining how much water has to be removed to achieve the desired concentration, we can estimate the membrane area required using the experimentally determined flux rates. As shown in Figure 4, these flux rates generally range from 8 to 12 gal/(day)(ft²) at 600 psi (for a medium-tight membrane and dilute effluents) to 3–7 gal/(day)(ft²) by the time the material has been concentrated to the desired final concentration. If a very high rejection level is desired and hence a tight membrane is selected, flux rates will be substantially lower and may range from 0.5 to 2 gal/(day)(ft²) at 600 psi (0.02–0.1 m³/(m²)(day) at 40 bar).

Preliminary experiments conducted on each type of waste reveal the minimum velocities that are needed in the module in order to avoid concentration polarization and to minimize fouling of the membranes. For tubular designs, a certain minimum flow rate has to be determined in each tube in order to obtain good performance.

Minimum required linear velocities that were determined experimentally for treating pulp mill effluents in 0.50-in. tubular RO systems are shown below:

Linear velocity	1–5 ft/sec (0.3–1.5 m/sec)
Flow rate/tube	0.66–3 gal/min (2.5–12 l/min)
Reynolds no.	4000–20,000

In general, values in the lower range of 1–2 ft/sec (0.3–0.6 m/sec) are valid for the lower concentrations in the range of 0.5–2% solids, whereas concentrations in the range of 5–10% solids often require velocities in the range of 3–5 ft/sec (1.0–1.5 m/sec).

The minimum linear velocity automatically sets a limit to the number of tubes or modules that can be put in parallel since the flow of effluent is limited. As the effluent moves through the reverse-osmosis system, water is

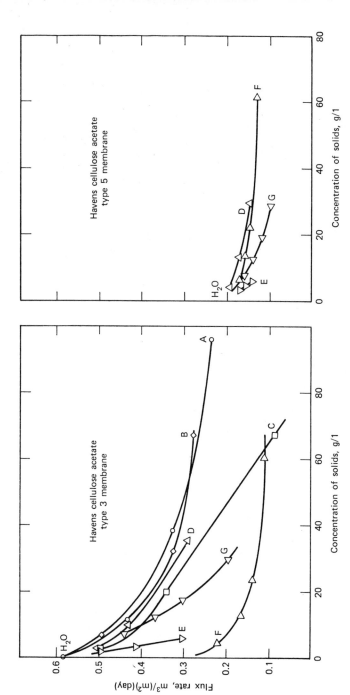

Figure 4. Flux rate at 25°C and 40 bars as a function of concentration for seven different waste waters. A, Diluted Ca-base spent sulfite liquor concentrate; B, Ca-base pulp wash water; C, kraft first-stage bleach effluent; D, kraft second-stage bleach effluent; E, sulfite first-stage bleach effluent; F, NSSC white water; G, evaporator condensate from NH_4-base liquor.

237

removed, which reduces the flow. Thus fewer and fewer modules can be used in parallel. The fact that the velocity requirements become more stringent at the higher concentration levels increases the necessity for fewer modules in parallel. For reverse-osmosis systems designed for concentrating purposes (such as the treatment of pulp-mill effluents), a number of tubes or modules must be put in series. For example, a system designed for concentrating an effluent ten times on a straight-through basis at an average minimum velocity of 3 ft/sec (1 m/sec) and a flux rate of 0.3 $m^3/(m^2)(day)$ will require 2050 m of 0.50-in. membrane-lined tubes in series to achieve tenfold concentration and to meet the velocity requirements. As a result of having so many tubes in series, the pressure drop is another factor that must be considered when designing reverse-osmosis systems for concentrating purposes.

E. Pressure Drop

The pressure drop in a reverse-osmosis system depends on a number of factors. First, it depends on the velocity (U) in the modules. For a commercial Havens module with eighteen 0.50-in. tubes in series with a total tube length of 39.6 m and a membrane area of 1.6 m^2, the hydraulic resistance as a function of velocity can be formulated as follows:

$$\Delta P = 1.4 U^{1.85}$$

where ΔP = hydraulic resistance (bar)
       U = linear velocity (m/sec)

With such a reverse-osmosis unit, the total pressure drop resulting from operation at a velocity of 1 m/sec is about 1000 psi ($\sim$70 bars). From this representative example, it is evident that reverse-osmosis concentrating units must be equipped with in-line booster pumps in most instances, or alternatively they must be equipped with pumps built into recycle loops to compensate for the pressure losses. For large reverse-osmosis plants with a limited concentrating effect, in-line booster pumps are more economical; for smaller units designed for substantial concentrating, recycle loops are the best design.

The number of modules in series needed for a given concentrating effect is directly proportional to the applied linear velocity. This factor, combined with the pressure drop–velocity relationship, leads to the conclusion that the total pressure drop in a system is approximately proportional to the square of the linear velocity. Accurate experimental determinations of the minimum required velocity are necessary to achieve an economical design for reverse-osmosis concentrating systems. For example, in the absence of experimental data, the operating velocity might be chosen to be twice the velocity actually needed. This doubling of velocity would increase the pressure drop about four times.

Another factor that affects the pressure drop is the flux rate. The higher the flux rate, the less holding time is required for the feed in the reverse-osmosis system to achieve the desired concentration. This results in reduced need for a number of modules in series. All these factors lead to the following results for pressure drop in a reverse-osmosis system:

$$\Delta P \sim \frac{U^2}{J}$$

where the terms have been previously defined.

V. DESIGN OF A FIELD DEMONSTRATION UNIT

The Federal Water Quality Administration (FWQA) and its successor, The Environmental Protection Agency (EPA), have supported research at the Institute of Paper Chemistry, Appleton, Wisconsin, to design, construct, and operate a 50,000–100,000 gal/day (190–380 m³/day) mobile reverse-osmosis unit. The main purpose of the research was to demonstrate, on a reasonably large scale, the utility of the reverse-osmosis process to the treatment of wastes in the pulp and paper industry. Besides the capacity to handle up to 100,000 gal/day, other main design criteria included the capability of concentrating dilute mill effluents with approximately 1 % solids to concentrations in the range of 5–15 % solids, and the capability of retaining most of the biochemical oxygen demand (BOD) in the concentrate.

A. Process Design

Pulp- and paper-mill effluents usually contain suspended solids and fibers, and this, as previously stated, was the reason for preferring a tubular reverse-osmosis design. Permeability of cellulose acetate membranes for both water and dissolved solids can be varied over a wide range by using different curing temperatures. This means that membranes can be obtained ranging from membranes with high flux rates and low rejection to membranes with low flux rates and high rejection. Guided by these general considerations, the Havens Model J, 18-tube module lined with Havens Type 3 membranes was selected as the best equipment available in 1968 for treating pulp wastes as based on the following specifications:

1. Proven capability of handling substantial amounts of fibers and suspended solids.
2. A fairly high flux rate of 15–20 gal/(day)(ft²) at 600 psi [0.6–0.8 m³/ (m²)(day)] for a test solution containing 5000 mg/l NaCl. Based on the results of laboratory experiments, this unit should have good flux rates [approximately 10 gal/(day)(ft²)] for dilute mill effluents.

3. A better than 90% rejection for total solids and a similar level of BOD rejection, provided the BOD was not largely caused by the presence of free acetic acid. Membranes now available all show poor rejection for low-molecular-weight organic acids, such as acetic and formic, and such acids may have to be neutralized to the salt forms if good BOD removal is required.

Because the purpose of the mobile unit was to demonstrate the potential of the reverse-osmosis process on several types of mill effluents, the minimum required linear velocities in the demonstration unit were designed on the basis of the highest velocities that were found to be needed for dilute mill effluents in previous laboratory experiments. The following assumptions for the minimum velocity were used:

Concentration	ft/sec	gal/min	m/sec	l/min
Less than 2% solids	2.5	1.5	0.75	6
2–6% solids	3.3	2.0	1.00	8
6–12% solids	4.2	2.5	1.25	10

An average flux rate of 7.5 gal/ft² [0.3 m³/(m²)(day)] was assumed for concentrating a 1% solids feed to a 10% solids concentrate. Straight-through operation under these conditions would permit only a small number of modules in parallel and would require a number of modules in series. This would result in a pressure loss of about 1200 psi (82 bars). This problem could have been solved by adding twelve 100 psi in-line booster pumps, but this was economically impractical, so a system was designed that could be operated with a minimum of two or three recycle loops. In this manner, a system was obtained in which the pressure was at no point more than 100 psi (7 bars) below the maximum operating pressure.

Detailed information about the actual design configuration of the mobile (610 m²) unit is presented in Figure 5. The total membrane area of 6600 ft² (610 m²) was distributed over four stages. Stage 1 was designed to be operated on a straight-through basis, whereas Stages 2 and 3 were designed for recycle operation. Mode I in Figure 5 is used with high rates of feed and with high flux rates. In Mode I operation, Banks II and III of Stage 2 are operated on a straight-through basis with Pump A being used as an in-line booster pump to aid in satisfying minimum flow requirements. For pulp-plant wastes in which Mode I operation is not feasible, Banks II and III have to be operated on a recycle basis, as shown in Mode II.

The system could have been designed with only one or two recycle loops, but this would have limited the possibilities for obtaining samples and process information covering the entire range of wastes to be treated. The design chosen (with three recycle pumps) permitted samples to be taken and process data to be obtained for at least four distinct levels of concentrations.

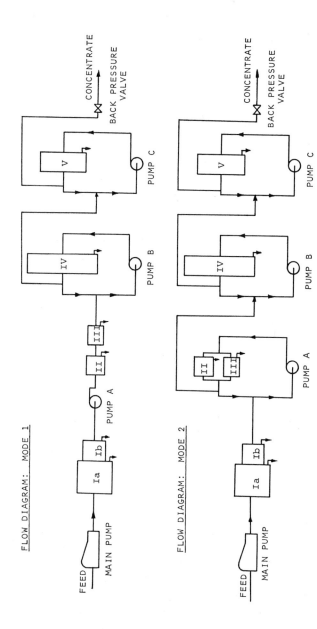

Figure 5. Flow diagrams for reverse-osmosis demonstration units to treat 380 m³/day of feed.

241

B. Hardware

The reverse-osmosis unit and accessory equipment was installed in a 40-ft over-the-road trailer so the test facility could be easily moved from site to site. Except for the initial supply of modules, all the hardware was designed for treating up to 100,000 gal/day (380 m³/day) at a maximum operating pressure of 1000 psi (70 bars). A total of 6580 ft² (610 m³) of Havens Model J, 18-tube modules were installed, which at a flux rate of 7.5 gal/(ft²)(day) [0.3 m³/(m²)(day)] would give a nominal capacity of 50,000 gal/day (190 m³/day). This capacity could be doubled by adding more modules. The maximum allowed pressure for the installed Havens modules was 600 psi (40 bar). The unit was equipped with safety switches, which automatically shut the unit down when the temperature, pH, or pressures were out of design limits.

The main pump was a triplex reciprocating Manton Gaulin pump with a 38-kW d-c motor and an electronic variable-speed drive. The recycle pumps are Goulds high-pressure centrifugal process pumps specially modified for a system pressure of 1000 psi (70 bars). The pumps were equipped with Crane single-balanced mechanical seals with a flush system for both the high- and the low-pressure sides of the seals. All metal components (e.g., pumps, piping, valving, and instrumentation) in contact with the process solution were 316 stainless steel.

Because it was found that for certain effluents a short period of zero pressure is effective in cleaning fouled membranes, pressure pulsing equipment was also installed. Pressure pulsing shuts off all the pumps periodically (e.g., each hour), which results in the pressure dropping to zero for a period of 1–5 min.

C. Performance

During a period of six months the unit was operated first on a calcium-base sulfite-pulp wash water and subsequently on a NSSC white water, which is a combined machine wash water from a neutral sulfite semichemical pulp and paperboard mill.

The average solids concentration of the calcium-base sulfite-pulp wash water was approximately 1 %. The calcium content varied from 500 to 700 mg/l. The pH ranged from 2.0 to 2.5 and was adjusted by adding about 6 lb NaOH/kgal (700 g/m³) of the pulp wash water. The larger fibers and suspended particles were removed with a 100 mesh/in. (150 μ) vibrating screen. Although this screening did not remove all suspended material, no problems were experienced with plugging of the tubes. A more detailed description of the composition of the pulp wash water is presented in Table 5.

Table 5. Analyses of Feeds to Reverse-Osmosis Test Unit

	Ca-base Sulfite Pulp Wash Water	NSSC White Water
Solids (g/l)	8–13	7–11
BOD_5 (g/l)	2.8–4.5	2–3
COD (g/l)	7–17	8–12
Volatile acids (g/l)	1.2–2	
Calcium (g/l)	0.4–0.7	
Sodium (g/l)		0.9–1.2
pH	3.5–4.5[a]	5.8–7.4

[a] After adjustment of pH.

During 690 hr of operation, the wash water that contained 1% solids was concentrated to 6–12% solids on an average.

During this period, a total of 1,260,000 gal (4800 m³) was concentrated and 45 tons of solids were recovered in the form of 6–12% concentrate, which was sent to the evaporators for further concentrating. As shown in Table 6, after 268 hr of operation it became apparent that Bank V, which had

Table 6. Summary of Flux Rate Date for the 190 m³/day Demonstration Unit; Pressures were 600–400 psi (41–27 bars)

Operating Period (hr)	Recovery of Water (Permeate) (%)	Concentration of Final Concentrate (% Solids)	Temperature (°C)	Flux Rate gal/(day)(ft²)	m³/(m²)(day)
Ca-Base Sulfite Pulp Wash Water					
0–100	90	12	23	6.9	0.28
100–200	85	9	24	5.2	0.21
200–268	90	12	26	4.8	0.19
Removal of $CaSO_4$ in Bank V with EDTA Solution					
268–400	83	6	28	6.2	0.25
400–500	80	6	27	5.7	0.23
500–587	78	5.5	28	5.7	0.23
Wash-up with Detergent for Removing Wood Pitch					
587–690	82	6	28	6.4	0.26
NSSC White Water					
690–790	80	5	33	5.9	0.24
790–1013	85	6	33	5.9	0.24
1013–1161	80	5	34	5.1	0.21
Addition of Bactericide to Feed					
1161–1317	87	7	35	7.2	0.30

been operating at the high final concentration, was scaled up with $CaSO_4$. This was removed with an EDTA solution and Bank V was subsequently put back in service. From here on, the maximum concentration was limited to 7–8 % total solids because this seemed to be a critical concentration for the formation of calcium scale.

Some experimental work was performed to explore the use of polyphosphates for the prevention of calcium scale. Results were promising and it seems likely that conditions will be found where Ca-base pulp wash waters can be concentrated to levels in excess of 8 % solids without scaling problems.

During the latter part of this first period it was found that wood pitch was fouling the membranes, particularly the modules of Banks I and II, which are at the inlet side of the reverse-osmosis unit. After 587 hr of operation, the wood pitch was removed with a nonionic detergent solution. The removal was easily accomplished, and the cleaning of the membranes substantially increased the flux rate.

As shown in Table 7, recoveries were excellent for the NSSC white water, resulting in a clear permeate of high quality. Also the BOD retention was almost as high as the retention of total solids because there were few, if any, low-molecular-weight organic acids in the NSSC white water.

For the calcium-base sulfite-pulp wash water, the recovery of values in the form of solids in the pulp wash water was good. The retention of BOD_5 and other pollutant materials was good, but not as good as for the NSSC feed. The retention of low-molecular-weight volatile acids was found to be poor when treating those liquors in which these components were present in high concentrations. Liquors from pulping of hardwoods may have high levels of acetic and formic acids. In addition, sugars may be biologically converted to these low-molecular-weight acids if liquors are recycled or stored for prolonged periods under conditions favorable for microbiological reactions.

Table 7.   Recovery of Material in the Reverse-Osmosis Demonstration Tests

	Ca-base Sulfite Pulp Wash Water	NSSC White Water
Solids (%)	85–90	95–97
BOD_5 (%)	70–80	88–93
COD (%)	85–90	95–97
Volatile acids (%)	25–40	
Calcium (%)	96–98	
Sodium (%)		87–94
Color (%)	96–98	>99

The permeation of acetic acid through the membranes affected BOD rejection because acetic acid contributes to the BOD. Conditions that minimize the formation of acetic acid were found to improve rejections and BOD removal.

The second demonstration was a NSSC white water, which is an approximately neutral wash water with an average total solids concentration of 0.9% (see Table 5). In addition to pretreating the feed with the vibrating 150-μ screen, the white water was run through a cooler to reduce the temperature to acceptable levels for the membranes used. No pH adjustments were made.

The demonstration on NSSC white water also proved the capability of the reverse-osmosis process to concentrate dilute waste effluents. For short periods, the concentrations recovered were as high as 10%. Operations were maintained for about 620 hr at feed rates of 40,000–50,000 gal/day (150 to 190 m³/day) (see Table 6).

It was evident that a considerable fouling of membranes was occurring and that velocities in the tubes were critical. After about 400 hr of operation on the white water, it became apparent that substantial slime growth was present in the reverse-osmosis system, due to the recycling and high operating temperature. An experiment in which a bactericide was added to the feed during the last 150 hr of operation showed an appreciable improvement in the flux rate.

VI. CONCLUSIONS

At this early state of its development as a new chemical engineering unit operation, reverse osmosis has been demonstrated to hold a high degree of promise for recovering high-quality water and for concentration processing the more dilute wastes from pulp and paper plants. Three specific areas favorable to the application of reverse osmosis for treating wastes from pulp and paper plants are (a) the recovery of values in the form of organic wood chemicals and of inorganic pulping and bleaching chemicals, (b) the removal of pollutants, and (c) the recovery of reusable water for recycle to the pulp and paper manufacturing operations. Because new developments of reverse-osmosis equipment and membranes are still being made rapidly and because of incomplete knowledge of module life and cost of replacement, the capital and operating charges to be expected for large-scale industrial installations cannot be predicted accurately, as yet. However, it is estimated that in the first years of commercialization, the operating charges will probably range upward from $0.50/kgal of recovered water permeate, and that costs at the present stage of development of commercial membrane modules are substantially higher than that estimate.

Field-scale research and demonstration studies, such as those for the co-operative project with the Federal Water Quality Administration of the United States Department of Interior, indicate that the first commercial applications will be those in which the costs can be justified by recovery of products of sufficient value to balance the relatively high processing charges until further improvements of membranes and equipment are made.

The research performed so far indicates that the applications for reverse osmosis in the pulp and paper industry are likely to be those in which dilute wastes (about 1% solids) are concentrated to a level of about 10% solids, so that conventional methods of evaporation could be used for further concentration to obtain usable products or to aid in final disposal of wastes.

For most dissolved materials in pulp and paper industry effluents, the rejections and recovery of dissolved materials may be expected to range upward from 90%. Most color bodies are removed at the 98–99% level. Exceptions to these excellent levels of rejection occur when the feed contains low-molecular-weight acids, such as acetic acid. However, it may be possible to minimize the presence of such materials in the dilute wastes to be treated by control of the formation of these acids during pulping and by inhibiting chemical and fermentation reactions that can occur during prolonged periods of storage and that form such acids.

Reverse osmosis of pulp and paper waste streams can be practiced with a minimum of pretreatment. The only pretreatments needed are (a) screening to remove large fibers, (b) adjustment of pH to values in the range of 3.5–7.5, and (c) adjustment of temperature within the range of 30–40°C. New membranes under development may further reduce the requirements for pretreatment to permit operation over wider ranges of pH and temperature than those specified for the present state of development.

The fluxes of permeate to be expected when treating pulp and paper dilute wastes is about 7.5 gal/(day)(ft²) [0.3 m³/(m²)(day)] for the 0.50-in. tubular configurations used in the studies described here. Extensive studies have indicated that such a tubular configuration is the best for processing pulp and paper wastes that contain small amounts of suspended solids in the feed or that form precipitates or develop microbiological slimes during reverse-osmosis treatment.

A critical factor in the design of reverse-osmosis systems for treating pulp wastes is the hydraulic pressure drop needed to maintain the minimum linear velocities through the tubes connected in series that have been found to be necessary to avoid concentration polarization and to minimize fouling of membranes. The requirements for pumping energy to overcome the pressure drop must be balanced against the disadvantages of prolonged holding time in a recycle system.

LIST OF SYMBOLS

C Concentration in solution above given membrane

C_c Concentration in final concentrate stream

C_f Concentration in feed stream

C_p Concentration in permeate from given membrane

$\bar{C}_p$ Average permeate concentration

F_c Flow rate of final concentrate stream

F_f Flow rate of feed stream

J Flux rate

Δ_p Pressure drop

$\bar{R}$ Recovery fraction of solute in concentrated stream $= F_c C_c / F_f C_f$

SR   Solute rejection fraction $= 1 - (C_p/C)$

U Linear velocity

REFERENCES

1. A. J. Wiley, A. C. F. Ammerlaan, and G. A. Dubey, *Tappi* **50** (9), 455–460 (September 1967).
2. A. C. F. Ammerlaan, B. F. Lueck, and A. J. Wiley, *Tappi* **52** (1), 118–122 (January 1969).
3. A. C. F. Ammerlaan and A. J. Wiley, "Water," *Chem. Eng. Prog. Symp. Ser.* **65** (97), 148–155 (1969).
4. A. J. Wiley, G. A. Dubey, J. M. Holderby, and A. C. F. Ammerlaan, *J. Water Pollution Control* **42** (8), Part 2, R279–R289 (August 1970).
5. W. R. Nelson and G. O. Walraven, *Pulp & Paper* **42** (34), 30, 31, 48 (August 19, 1968).
6. *Lockwood's Directory of the Paper and Allied Trades*, Lockwood Publishing Company, Inc., New York, 1969, p. 11.
7. D. C. Haynes, (Project and Disposal Committee Assignment 9015) *Tappi* **49** (9), 52A (September 1966).

Chapter XII   The Treatment of Industrial Wastes by Pressure-Driven Membrane Processes

Robert W. Okey*

I. INTRODUCTION

The water-borne contaminating or polluting materials that arise as a result of industrial operations are called industrial or trade wastes. Because these waste streams often contain a number of complex materials, they offer a perplexing and difficult problem to the designer confronted with the task of developing processes for their treatment. Although some wastes are easily treatable by well-known and well-understood processes, others often present real problems because of their chemical or physical character. The essential point is that industrial wastes may present the designer of treatment processes with many special problems beyond those faced in the treatment of domestic sewage.

The most important single deficiency in conventional treatment systems is the inability of these systems to make one-step separations on a molecular or ionic level. Conventional systems may employ processes based on biological conversion or a special chemical affinity, but in each case, a special intermediate step or process is required to remove the desired species or waste component.

The ability of pressure-driven membranes to make one-step separations on a molecular or ionic level and to make a further differentiation on the basis of charge and molecular character, under certain conditions, is thus of

* Resource Engineering Associates, Inc., Stamford, Connecticut.

pertinence to engineers concerned with the treatment of industrial wastes. This chapter contains a discussion of how this tool may be employed to increase the efficacy of industrial waste treatment processes or how it can replace conventional operations.

II. THE NATURE OF INDUSTRIAL WASTES AND THE EFFECTS OF WASTES ON RECEIVING WATERS

Industrial wastes may be categorized in several fashions. For present purposes, general categories that relate functionally to the use of pressure-driven membrane systems will be established. In Table 1, the principal

Table 1. General Character of Industrial Wastes

Industry	General Makeup of Wastes
Food	
Meat	Organic—very strong,[a] high in soluble, colloidal, and particulate matter
Vegetable and fruit	Organic—weak[a] to very strong, generally high in sugars
Bakery	Organic—weak to medium[a]
Chemical	
Refining, cracking, and petrochemical	Organic and inorganic—many small and large molecules, oils, sulfur compounds
Fiber	Organic and inorganic—small molecules
Fine chemical	Organic—frequently toxic or bacteriostatic, wide range of molecular sizes
Fertilizer	Organic and inorganic—may be very high in urea or ammonia
Pulp and paper	Organic, primarily (varies with process); can be high in lignin-related compounds and sugars, and fibers
Metals	
Iron and steel	Inorganic—high in ferrous and ferric sulfates from scale removal operations, acid
Plating	Inorganic—toxic heavy metals and cyanide
Agricultural	
Feed lots and dairy operations	Organic and inorganic—high in oxygen-demanding material and in algal nutrients, nitrogen, and phosphorus

[a] Very strong, BOD >1000 mg/l; medium, BOD 200–1000 mg/l; weak, BOD <200 mg/l.

industries and general information concerning the physical, chemical, and biochemical features of their wastes are presented. It is also possible to categorize flow data for each industrial group and subgroup based on the specific production process and the character of other water-using operations. Wastes from these industries have an impact on receiving waters in several ways. A brief discussion of these effects is a necessary prerequisite for an analysis of the applicability of pressure-driven membrane processes to the treatment of trade wastes.

The most commonly observed effect on receiving waters is that of oxygen depletion associated with the discharge of organic materials that serve as carbon and energy sources for saprophytic bacteria. These organisms do not use atmospheric oxygen directly but withdraw the gas from solution. The solubility of oxygen is limited to less than 10 mg/l under most conditions. Hence comparatively small discharges of organic material can create conditions of low dissolved oxygen or even anaerobic conditions on some occasions. Oxygen-consuming wastes are probably the most important category of wastes because of the large number of industries discharging great volumes of oxygen-consuming materials. High-oxygen-demanding wastes come from the pulp and paper, food, and chemical industries.

A second and important category includes those wastes which, although organic in character, are also toxic to the biota of receiving waters. These materials also create real problems where surface waters are used as a domestic water supply. Such materials can, and often do, persist for great distances and long periods.

A third category of major importance includes heavy metals that arise from metal-plating operations. The metals most commonly encountered are chromium (both 6^+ and 3^+), copper, cadmium, nickel, and zinc. These wastes may also contain substantial quantities of cyanide and are a problem because of their extremely high toxicity to receiving water biota. Heavy metals derive their toxicity from a substantially irreversible reaction with the vital sulfhydral groups contained in enzymes and other proteins. The toxicity of the metals is generally proportional to the insolubility of the corresponding sulfide.

Many substances not possessing marked toxicity can persist for substantial periods in receiving waters. The importance of these materials in receiving waters is often not known. However, some workers consider the presence of such biorefractory materials a potential hazard. For this reason removal of biorefractory material is sometimes required.

Other wastes can bring about physical or chemical changes which can make the marine or aquatic environment unfit or the receiving water unsuitable for domestic or recreational purposes. These wastes alter the pH, affect the color or turbidity, or cause the deposition of materials in sludge beds.

Hence, in summary, most industrial wastes may be broadly classified into five major categories:

1. Oxygen-consuming organic materials.
2. Toxic organics.

3. Toxic inorganics.
4. Biorefractory substances.
5. Substances that bring about general chemical or physical alterations of receiving waters.

III. CURRENT INDUSTRIAL WASTE TREATMENT PRACTICES

When a designer is evaluating an industrial waste treatment problem, he considers several important aspects in addition to the treatment of the waste stream. These additional considerations permit the selection of the optimal processes and indicate any in-house changes or controls that may be necessary to minimize overall cost.

Important parts of an industrial waste study are as follows:

1. An appraisal of the industrial processes including such aspects as (a) water use and (b) lost product.
2. Consideration of process modification to reduce loss or to reduce waste volume and strength.
3. An analysis of effluent requirements that relate specifically to the materials contained in the wastes.
4. A coevaluation of possible treatment processes and product recovery techniques, considering (a) effluent quality and (b) capital costs; usually amortized over 3–5 yr periods depending on the industry and equipment.
5. Operating costs.

In addition, many industrial wastes are complex enough to require special treatability studies in which a careful evaluation of the flow sheet proposed is carried out. Such basic factors as flocculation and sedimentation rates, rates of biochemical conversion, and other factors are determined.

The division made earlier based on the wastes' organic or inorganic nature may be rationally employed when normal or conventional treatment procedures are discussed. The treatment of industrial wastes follows patterns dictated by their physical and chemical makeup. These patterns are shown in a schematic fashion in Figures 1 and 2.

As can be seen, most organic wastes are treated by the removal of floatables and settleables, either by natural or aided sedimentation or flotation. The soluble materials are converted to biological cellular material, usually by employing aerobic systems. The cellular material thus produced is separated by gravity and disposed of by thermal, chemical, or biochemical oxidation.

Most inorganic wastes of a metallic nature are removed as the hydroxides.

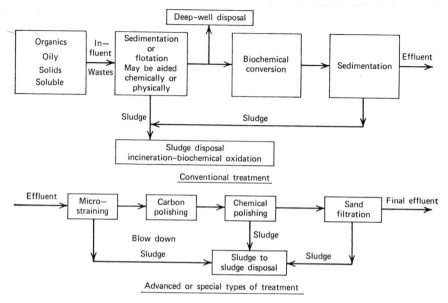

Figure 1. Treatment processes for organic wastes.

Those metals that do not form insoluble hydroxides or some toxic anions are either not treated, or treated by evaporation or deep-well disposal techniques.

The capital and operating costs of the commonly employed techniques are shown in Table 2. There is a substantial cost benefit from scale to be derived from the employment of conventional processes. For that reason, Table 2 contains costs and operating data for several different plant sizes.

The costs of several special treatment techniques that are frequently employed to improve effluent quality are shown in Table 3.

IV. PRESSURE-DRIVEN MEMBRANE SYSTEMS FOR TREATING INDUSTRIAL WASTES

Pressure-driven membrane systems remove species contained in true solution or colloidal suspension by two mechanisms: by ultrafiltration, in which water moves through pores, or discrete holes in the filtering media, and solute rejection is based primarily on size; and by reverse osmosis, a physicochemical process in which the species in solution is rejected as a result of its chemical characteristics rather than its size, which may be of the order of water. In the latter case, the aqueous phase apparently is transported

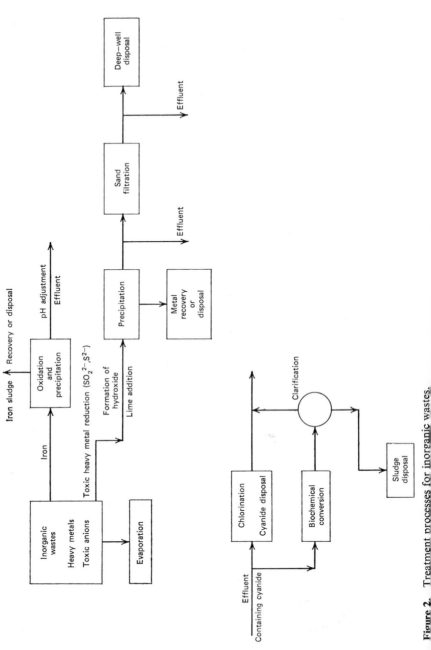

Figure 2. Treatment processes for inorganic wastes.

254

Table 2. Capital and Operational Cost Ranges for Conventional Industrial Waste Treatment Facilities[1-3]

Process or Operation	Capital Costs ($/10⁶ gal)			Operating Costs ($/10⁶ gal)
	<1.0 × 10⁶ gal/day	1–3 × 10⁶ gal/day	3–5 × 10⁶ gal/day	
Chemical treatment				
Neutralization or coagulant flocculation	50,000	42,000	32,000	10–50
Sedimentation	90,000	75,000	60,000	10–100
Biological conversion[a]				
Activated sludge	700,000	300,000	200,000	10–2000
Aerated lagoons	500,000	250,000	150,000	10–2000

[a] Includes sludge disposal—about 20% of total capital costs.

through the film by making and breaking chemical bonds with specific functional groups in the film. A pressure difference provides the energy to transport the water molecule. The solute, on the other hand, is largely insoluble in the water-swollen membrane or diffuses through it very slowly. Thus, the relationship between the membrane and the chemical characteristics and sizes of the species contained in industrial waste is of importance in choosing the type of membrane to be used.

In general, if molecules being rejected are 10 times the size of water, or greater, the process is termed ultrafiltration but the effective size of a molecule due to its hydrated sheath is hard to determine. As a consequence, the line

Table 3. Cost of Special Separation or Disposal Techniques Employed in Industrial Waste Treatment[3, 4]

Process or Operation	Capital Costs ($/10⁶ gal)			Operating Costs ($/10⁶ gal)
	<1.0 × 10⁶ gal/day	1–3 × 10⁶ gal/day	3–5 × 10⁶ gal/day	
Activated carbon	2.0 × 10⁶	1.5 × 10⁶	1.0 × 10⁶	80–200
Microfiltration	50,000	47,000	45,000	50–200
Evaporation	1.5 × 10⁶	1.1 × 10⁶	0.9 × 10⁶	500–3000
Foam separation	70,000	50,000	40,000	30–100
Deep-well disposal	1,400,000	1,200,000	1,000,000	250–400
Ion exchange	400,000	350,000	300,000	200–1000

Table 4.   Details of Species to be Removed from Industrial Wastes

Waste Source	Species	Molecular or Atomic Weight	Applicable Membrane Process
Iron and steel,	Fe^{2+}, Fe^{3+}, SO_4^{2-}	<100	Reverse osmosis
with pH correction	$Fe(OH)_2$, $Fe(OH)_3$	Colloidal or particulate	Ultrafiltration
Metal plating,	Mn^{2+}, Cu^{2+}, Cd^{2+}, Pb^{2+}, Cr^{3+}, and Cr^{6+}	<225	Reverse osmosis
with pH correction	Corresponding hydroxide	Colloidal or particulate	Ultrafiltration
Food processing			
Meat	Blood constituent and paunch manure	Inorganics <100 Organics 150– >10,000	Reverse osmosis Reverse osmosis or ultrafiltration
Food and vegetable	Sugars	100–300	Reverse osmosis
processing waste	Protein	5000	Ultrafiltration
Pulp and paper	Inorganics	100–300	Reverse osmosis
	Sugars	150–2000	Reverse osmosis
	Lignins and lignin sulfonates	>1000	Ultrafiltration
Chemical and	Small molecules	100–1000	Reverse osmosis
petrochemical	Monomers, dimers, etc.		Reverse osmosis
	Oils	Particulate or emulsion	Ultrafiltration
Any organics after biological conversion	Bacterial cell tissue	Particulate	Ultrafiltration

of demarcation between processes is indistinct. The species or materials to be removed with and without integration with other types of systems are presented in Table 4.

The data presented in Table 4 can be used as a base in generalizing on the separatory operations required for the important industrial contaminants. The principal contaminants fall into four categories:

1. Ionic species.
2. Organic in true solution.
3. Colloidal materials.
4. Suspended or particulate solids.

From the foregoing, it is possible to categorize the specific membrane processes that might be of use in treating industrial wastes in terms of conventionally defined membrane operations.

1. Removal of ionic species. Reverse-osmosis processes that are similar to those used for desalination of brackish waters may be used for metal or toxic anion removal.

2. Removal of small organic molecules. These operations may be carried out with reverse-osmosis membranes or, in some instances, with ultra-filtration membranes. In either event, the primary mode of membrane transport required for good separation is facilitated transport or simply diffusion.

3. Removal of colloidal and larger particles. Ultrafiltration processes employing films with the capability to retain particles above 5000–30,000 MW have been employed for this application.

Because of the nature of most industrial operations and their wastes, the waste stream carries the more hydrophilic of the materials processed. Therefore, if the principal modes of solubilization (hydrogen bonding and ionization) are considered, it must be concluded that most soluble industrial waste components contain substantial numbers of functional groups that aid in solubilization. These components, if small (i.e., of the order of water), tend to pass through the membranes now in use. Therefore, all things considered, size may be the single most important criterion in evaluating the suitability of membrane processes for these waste streams. As will be seen, however, other molecular characteristics can be used under certain conditions to effect separations and should also be considered.

A. Operating Characteristics

1. Membrane Fouling

Many wastes have substantial quantities of suspended solids and these wastes impose certain special restrictions or constraints on a membrane waste-treatment process, particularly with regard to the spacing between films, to the superficial liquid velocity, and to the need for avoiding bends or dead spots in the flow channels. The most important need is to provide conditions at the membrane surfaces that minimize the effects of deposition in the boundary layers adjacent to the surfaces.

Dorr-Oliver workers[5] have studied the problem of surface deposition and means of minimizing its effects on flux. This work with the Dorr-Oliver multiplate membrane module demonstrated that the superficial velocity of the solution was the primary regulator of stable flux over a wide pressure and velocity range. Although not proved, the effects of velocity on regulating deposition probably relate to the thickness of the boundary layer and the ability for ionic and colloidal materials to diffuse away from the membrane surfaces. The thickness of the boundary layers vary with the Reynolds number of the solution flow past the membranes. In a fixed configuration of equipment treating a fluid with uniform properties, the thickness of the boundary layer at a given point is governed by velocity only.[5, 6]

The foregoing leads to some conclusions as to how the film should be held. Since there is flow out of the forming boundary layer, it could be argued that a system in which the flow was broken *before* the stable layer is

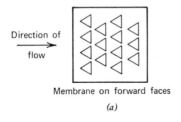

Membrane on forward faces

(a)

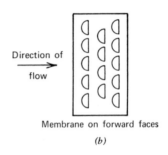

Membrane on forward faces

(b)

Figure 3.   Possible stack configurations.

formed would be desirable. Also, the use of mechanisms to break up dead areas may be useful. The impact on turbulent diffusive transport in an actual system probably is controlled by the frequency of the surface perturbations. However, little investigation of these aspects has been made. It is becoming increasingly apparent that such work is urgently needed.

The concept of controlling the formation of the stable boundary layer leads to film-support configurations not observed in literature reports. These stack concepts are shown in Figure 3. The purpose of such a configuration is two-fold: first, to provide an impinging flow, and secondly, to break the flow before the stable boundary layer is formed.

Systems do not exist now that can take full advantage of the available high-flux films developed by the Amicon Corporation and others.[7, 8] This lack is retarding the orderly application of membrane systems. Much attention has been devoted to the membrane itself, and very little to the device for holding or supporting it. Consider the possibilities associated with realizing 150–200 gal/(ft²)(day)—occasionally 50–75% of this clear water flux of the open ultrafiltration film is realized, but in most instances, only 5–10% is realized.

Since high velocity (>3 ft/sec) or high Reynolds number (>3000) appears to be required when handling nonhomogeneous waste streams, certain system configurations are indicated. If an adequate velocity is to be maintained across the surface of film in a stack of practical size, a substantial recycle rate is required. Depending on the stack configuration, recycle rates of 10–50 to 1 may be required. A typical flow sheet is shown in Figure 4.

The configurations involving a high degree of recycle have an operating constraint associated with the concentration of materials in the recycle stream. With inorganics, the constraint is osmotic pressure and well understood. Well understood also are the effects of suspended material in the recycle line and the impact of these solids on operations. Work on activated sludge has shown that 20,000–30,000 mg/l of biological solids can be carried within the recycle loop. At concentrations in excess of 30,000 mg/l, marked diminution in flux was noted. However, the impact on flux is probably related to the character of the solids and each case should be examined in this regard.[9]

Membrane systems are available commercially in at least three basic forms: (a) the porous-plate form offered by the Dorr-Oliver Company; (b) the tubular design offered by Gulf General Atomic, Inc. and Havens Industries; and (c) the spiral-wound design offered by Aerojet General Corporation. Some partially successful attempts have been made to insert flow perturbators in the tubular system. However, all these systems suffer from the same basic problem; they all permit the formation and persistence of a stable laminar layer or sublayer, and for waste-management applications there would appear to be little choice between them except on the basis of cost and pressure limitations. A new and fresh approach is needed.

2. Pressure Requirements

Great variations in the concentrations of materials in wastes can be anticipated. In most cases, the dissolved solids content of the waste will be low, but metal-plating waste is sometimes an exception to this generalization. In general, however, osmotic pressures do not exceed a few psi and the osmotic pressure across the membranes is usually of minor importance in

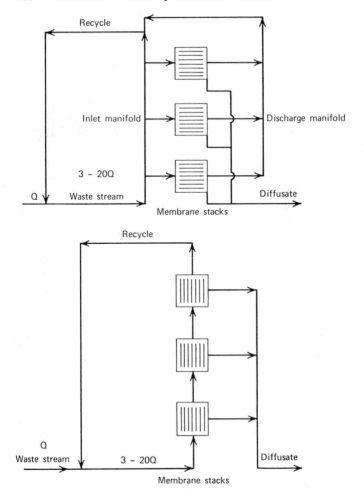

Figure 4. System configurations. (Q represents the volumetric flow rate of the waste stream.)

choosing the hydraulic pressure to be used. The hydraulic pressure is usually chosen to supply the desired transport rate through the membranes and the pressure drop required to maintain the desired superficial velocities.

Systems treating organic waste streams or systems integrated with biological conversion systems usually utilize pressures less than 200 psi and frequently less than 50 psi. Since osmotic pressure is a direct function of molarity, relatively high concentrations of most organic wastes produce only low osmotic differentials. For example, the osmotic pressure of a 45,000 mg/l (4.5%) sucrose solution is 3.14 atm or less than 50 psi at 20°C. A 2-M

solution of cadmium cyanide (3.2%) has an osmotic pressure of 70 psi. Therefore, although some applications are akin to desalination, the actual osmotic pressures are much lower than normally encountered in desalination operations, because of major differences in molecular weights between the heavy metal salts and sodium chloride and other salts in natural waters to be desalted. This factor renders pressure-driven membrane systems attractive for dewatering or concentrating waste components of high molecular or atomic weight because of the relatively low hydraulic pressures required.

B.   Membrane Treatment of Organic Wastes

As outlined earlier, most nonsettleable organic wastes are conventionally treated by the conversion of the organic material to bacterial cell tissue. With careful design and operation, biological conversion facilities and their accompanying clarifier can reduce the total BOD to 20–50 mg/l in the effluent. Although this is usually an effluent of satisfactory quality, there are times when a higher-quality effluent is desired, and additional safeguards against short circuiting or effluent polishing may be required. The pressure-driven membrane system may be employed in a variety of ways to treat organic wastes or polish the effluents treated by other means. The technique to be employed is governed by the character and quantity of the pollutants.

1.   Direct Waste Treatment

The obvious waste-management or product-recovery technique is to treat the unaltered or screened waste directly. Indeed, it is likely that where a limited number of components is contained in the waste stream, such an operation is possible. Frequently, the waste contains a sufficient spectrum of molecular sizes, charges, or both, to make the selection of a nonfouling film with sufficient retention difficult, if not impossible. The work of the Pulp Manufacturers' Research League[7, 8] on kraft liquor (discussed in Chapter 11), the work of Dorr-Oliver,[9] and the reports of Bixler[10] indicate that proper membrane selection is difficult under these circumstances. Yet these studies show that on a single well-defined component, or when two components with substantial size differences are present, retention or separation at low pressure is possible and direct treatment is possible.

The direct-treatment procedure appears best suited to those cases in which the primary species to be retained is high in molecular weight. Organics larger than 1000 MW—and preferably above 10,000—fall into this category. Such a system can be employed for stripping an organic from a waste stream containing a noncontaminating concentration of inorganics.

Where the waste is more complex, as is often the case, direct treatment is not practical, or indeed possible. Therefore, some special means of pretreatment or preconditioning is required. In addition, some pretreatment techniques can be used as an aid in making separations at higher flux values than normally possible, as discussed in the following section.

2. Biomembrane Systems

One of the most thoroughly examined membrane techniques for the treatment of organic waste is the integration of biological conversion techniques with pressure-driven membranes. In this case, the membrane serves to retain the biological material or the biological material and the unassimilated substrate.[13]

The basic flow sheet employed is comparable to bench scale systems employed by Dorr-Oliver,[13] Amicon,[10] and others.[11] The concept incorporates the ability to recycle varying amounts of flow for the purpose of scrubbing the fouling film from the membrane surface. The flow sheet is shown in Figure 5.

The membrane system employing biological conversion has many desirable features:

1. Large unassimilated organics may be retained by the membrane.
2. Slowly metabolized organics may be retained by the membranes to

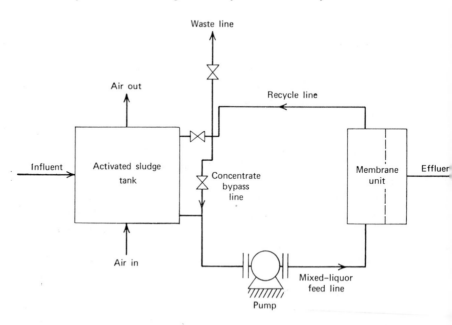

Figure 5. Waste treatment flow sheet.

increase their effective contact time within the system without a corresponding increase in the hydraulic size of the facility.

3. Hindered settling, the factor limiting the concentration of biological material in conventional treatment, is not a consideration (i.e., a very high solids concentration may be maintained in the biological reactor). The required detention time may be reduced or the treatment efficiency at the same contact time may be increased.

Although the Dorr-Oliver systems have been operated on sanitary sewage rather than on industrial waste, the performance of these systems provides some insight into the operation of biomembrane systems.

Pertinent operating data are presented in Table 5.

Table 5.   Performance of Biomembrane Plants Treating Sewage or Effluents[13]

Item	Value
Membranes	Ultrafilters
Flow	3600 gal/day raw sewage
	20,000 gal/day on secondary effluent
Flux	8–12 gal/(day)(ft²)
	Pressure at 15–25 psi
Effluent	
Suspended solids	0
BOD	90% of the time = 14 mg/l
	50% of the time = 3 mg/l
Coliform	Geometric mean
	100 org
BOD reduction	95%
BOD reduction[a]	98%
COD reduction[a]	90%

[a] From bench scale pilot-plant data.

3.   Carbon as a Biomembrane System Adjunct

Small molecules may be retained in a biomembrane system otherwise not capable of capturing such species if carbon is used as a further adjunct to treatment. Although few data are available concerning this procedure, there are many possible and interesting applications.

The Dorr-Oliver workers[11] have used powdered carbon as a preconditioner for a biomembrane system containing no activated sludge. The carbon apparently serves to prevent fouling of the membranes with oils and greases.

The total solids in the complete system changes as biochemical conversion proceeds and apparently the carbon is regenerated biochemically. The work of the Dorr-Oliver workers demonstrated that retention increases more rapidly than do the biological solids in the system, and that normal flux is rapidly attained. Retention and surface protection are provided by carbon as could be expected from first principles. In general, the adsorption of organics on activated carbon follows Traube's law, which in simple terms is that the greater the lipophilicity of the molecules, the greater the extent of adsorption. Therefore, carbon should protect a film most specifically from those materials that would tend to increase surface fouling because of their water solubility.

Other considerations are also relevant to use of carbon as a preconditioner:

1. Small molecules possessing a marked difference in lipophilicity can be separated in high-flux, low-pressure systems.

2. Two or more molecules of equal lipophilicity but with different ionization properties can be separated employing this technique. For example, butanol or propanol and phenol probably are not separable at pH 7 but probably are at pH 10, above the pK for phenol.

3. The flow process through a series system may be organized to permit efficient carbon use. For example, countercurrent carbon flow may be practical in an effluent-to-loop series system or in a loop-to-loop scheme.

The greatest advantage of this concept, however, is the ability to employ open, high-flux membranes with a concomitant reduction in the size of the system. Membranes capable of clear water fluxes of 100–500 gal/(ft²)(day) at 100 psi are envisioned for this use.[5]

Probably the greatest drawback to the carbon-membrane scheme is the lack of reliable systems for carbon regeneration. Where recovery of the carbon is not important, direct incineration with heat recovery can be practical. The Dorr-Oliver concept of biochemical regeneration could be employed on biodegradable contaminates.

C. Membrane Treatment of Inorganic Wastes

The basic data that have been developed for the desalination of seawater or treatment of brackish water indicate that membrane systems may be profitably used in the treatment of a variety of waste streams that contain inorganic materials. Among the industries or operations to be considered are metal plating, nutrient chemical removal, water softening, and desalting of sewage effluents.

Although the waste-treatment problem is essentially one of the removal of ions, there are some techniques that can be used to facilitate the removal of the smaller species by increasing their size.

1. Metal-Plating Wastes

Metal-plating wastes are derived from the rinsing of plated materials after plating and result from the "drag out" of plating solution from the plating tank. The quantity and strength of the waste stream are a function of several system variables including the nature of the plated parts, the rate of plating and rinsing, and the strength of the plating solution.

Modern plating facilities consist of several plating lines, each of which has a plating tank and several rinse tanks. The fresh water enters the last rinse tank and flows through several (2–5) stages countercurrently to the materials being plated. It is discharged from the tank adjacent to the plating tank, as shown schematically in Figure 6.

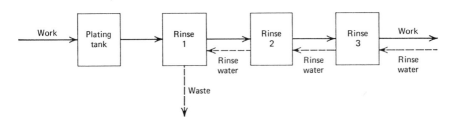

Figure 6. Typical plating line.

The concentration of metal in the plating tank might vary from 0.5 to 3.0%. The concentration of metal in the waste stream usually varies from 25 to 1000 mg/l. The flow rate of most rinse streams is from 400 to 2000 gal/hr. Other sources of waste are spillage from the plating tanks and accidental spills of chemicals during handling.

The metals most commonly encountered are as follows:

1. Chromium (hexavalent)
2. Copper
3. Cadmium
4. Zinc
5. Tin
6. Nickel
7. Lead
8. Silver
9. Gold

The first six metals are most commonly encountered. Many of these are usually plated out of a cyanide solution. Therefore, this toxic anion is commonly encountered in plating wastes. The presence of cyanide tends to dictate the pH of the processing system and the operating pH of any waste recovery or treatment system.

Conventional systems to treat plating wastes are often awkward, expensive, and generally unsatisfactory. There has been a tendency toward "ancestor worship" in the treatment of plating wastes that results in the choice of a method of waste treatment because it has been used for years instead of a cheaper, more efficient new method. Plating wastes have traditionally been treated by the following steps:

1. The reduction of Cr^{6+} to Cr^{3+} with SO_2 or S^{2-}.
2. The adjustment of pH to permit the formation of the hydrated oxide or hydroxide.
3. The flocculation of the hydrated oxides.
4. The precipitation of the hydrated oxides.
5. The chlorination of the effluent to convert the cyanide to cyanate or nitrogen and carbon dioxide.
6. The concentration or dewatering of the sludges.
7. The subsequent disposal of the sludge.

This method of treatment has several important disadvantages:

1. It requires close operational control for good performance.
2. Sludge disposal represents a major problem. Some states are considering prohibiting the disposal of metal hydroxides by spreading or dumping on the ground.
3. The necessary combining of waste streams makes recovery difficult, if not a practical impossibility.
4. The space required for the needed equipment is relatively large.

These factors present a real opportunity for a compact, effective system permitting a degree of return of both metals and rinse waters.

Previous work[10] has indicated that membranes with moderate sodium chloride retention (50–70%) will retain essentially all (>99%) of the polyvalent cations. A membrane-treatment process that appears promising is shown in Figure 7. The proposed membrane process has a great advantage over most other systems capable of treating plating wastes in that it is modular in makeup. This modular character permits a wide variety of physical configurations. Space is frequently a problem in plating works, and the ability to stack the membrane modules in a random or varied fashion is an extremely desirable attribute.

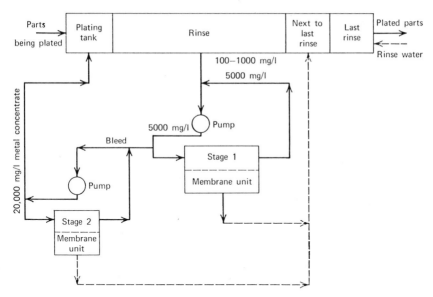

Figure 7. Possible metal recovery flow sheet.

The values of recovered products and the payout times of a membrane process to treat plating wastes are presented in Table 6. A total operating cost of $2.00/1000 gal has been assumed for the analysis in Table 6, although in the future, operating and maintenance costs should be reduced considerably below this. (Ultimately, the total costs of operating such a system should be between $1.00 and $1.50/1000 gal.) The bulk of the present operational costs, roughly $1.30/1000 gal, can be attributed to the cost of membrane replacement. With the developments now under way the costs of replacement of membranes should be markedly reduced. Still, considering most companies' depreciation schedules, the present costs for systems treating the more expensive metals should be attractive based on the payout time.

An important point to recognize is that changes in capital cost do not have a marked effect on system economics. This will be true as long as the operating costs continue to be so much larger than the amortized capital cost.

As with organics, process tricks may be used in treating metal-plating wastes. Mention was made earlier of the concept of increasing the size of the species to be removed. Metals may be complexed by a variety of organics. For example, EDTA and dithizone both form large (in a molecular sense) complexes that may be removed on a much more open film than would be possible with the ion. Some complexing agents, such as the foregoing, will form soluble complexes. Others, such as dimethylgloxime, form an insoluble

Table 6. Economic   Relationships   in   Treating   Metal-Plating Wastes

Basic Metal Cost ($/lb)	Value of Metal ($/day)	Value of Water ($)	Total Recovered Value		Payout Time (8%) (yr)
			($/day)	($/year)	
1	25	7.50	32.50	11,900	—
2	50	7.50	57.50	21,000	—
3	75	7.50	82.50	30,800	3.8
4	100	7.50	107.50	39,300	1.9
5	125	7.50	132.50	48,400	1.2
6	150	7.50	157.50	59,500	1

System flow 30,000 gal/day
Metal concentration 100 mg/l
Cost of water $0.25/1000 gal
Initial cost of membrane treatment plant $30,000 (flux 10–20 gal/(day)(ft^2))
Metal lost per day 25 lb
Operating cost $2.00/1000 gal (Annual = $21,900)
Net dollars applied to capital investment = total recovered $/yr − operational cost

$$\text{Payout time} = \frac{\text{Capital cost}}{\text{Annual net savings adjusted for interest requirement}}$$

complex with nickel. The oximes in general are useful for the purposes described in this paragraph.

Many of the complexing agents are highly pH-sensitive, which could permit full recovery and reuse of the complexing material after the release of the metal into the second low-flow system treating the concentrated wastes in ionic form.

One process trick that often may be employed to advantage in the treatment of metal-plating wastes is the precipitation of the metal as the sulfide. With such precipitates a much more open film may be employed. Recovery is more difficult with this procedure since the winning of the metal from the sulfide often is a difficult problem. Under some circumstances, however, this latter point may not be a problem. With some metals, the anions will be discharged through the films if alkaline correction is made with Na^+ or NH_4^+ ions. If the anion is cyanide, further treatment of the liquid stream will be required. Further treatment of the concentrate stream is also necessary. Metal sulfides are easily coagulated with polyelectrolytes and centrifuged to a high solids concentration (>30% solids).

2. Removal of Nonmetallic Inorganic Species

In water management, there are at least two areas in which removal of inorganics is important and for which membrane systems appear to offer the possibility of a substantial improvement over existing processes, either in efficiency or in actual costs. These two areas are the removal of specific inorganics from waste streams, and water softening. The latter application is perhaps beyond the scope of this chapter, but is mentioned here as an increasingly important environmental management application.

The removal of phosphorus, an important algal nutrient, may be accomplished with membrane systems. Okey et al.[6] report on the use of an integrated biomembrane system in which 22% of the influent phosphate was removed at a pH of 7–8. When the pH of the completely mixed material in the reactor was raised to 8.5–9.0, the removal of phosphate was around 90%. The membrane is apparently able to retain calcium phosphate in some form that would not normally be removed at the normal values of pH (7.0–8.0). This result may cast some doubt on the usual solubility data. This salt or complex form of calcium phosphate was retained by a very open film whereas 100% of the monovalent species and some small organics passed through the membrane.

Water may be softened by the direct removal of calcium and magnesium, the two metals usually responsible for hardness. The problem is closely akin to the problem of heavy metal removal. The membranes employed usually have sodium chloride rejections ranging from 50 to 80%, and 99% rejection of calcium and magnesium sulfate, chloride, and bicarbonate. Membrane systems may have a decided advantage over other methods of softening. Both the lime–soda ash and the ion-exchange processes produce a substantial quantity of waste. The lime–soda ash process produces a calcium carbonate and magnesium hydroxide sludge that can be dewatered only with considerable difficulty and that poses a major ultimate disposal problem. Ion exchange produces a high-solids stream during regeneration that may pose a disposal problem. The concentrate stream that results from softening water by reverse osmosis does not contain suspended material and can usually be disposed of more easily than the wastes from the other two processes.

Some attention has been directed in the past toward the treatment of acid mine drainage and irrigation return flows. These streams tend to be large in volume and contain waste materials of limited value. Hence although some of this work has been an unquestioned technical success, the cost of membrane treatment is such that the approach does not seem practical under present conditions.

D.   Miscellaneous Waste-Management Applications

There are several waste- or water-management operations that are either marginally feasible now or await substantial reductions in the fouling problems that increase both capital and operating costs. These applications are as follows:

1. Algae harvest from algae-growth units or sewage-polishing lagoons.
2. Colloid and particulate removal from water or wastes for reuse.

1. Algae Harvesting

One of the more perplexing problems in the tertiary treatment of sewage by lagooning is the problem of harvesting the algal cell tissue and the large quantity of extracellular organics produced by algae.

A truism about such lagoon treatment (or any treatment involving the growth of cells) is that unless the cell material is removed from the system, little permanent good is accomplished. The effluent nutrients eventually come into steady state with the influent materials unless some removal is accomplished. A variety of techniques for removing cell material have been attempted with only limited success. Among the procedures tried are the following:

1. Coagulation and sedimentation or centrifugation.
2. Flotation with and without coagulants.
3. Sand filtration.

Straight coagulation and sedimentation may be the least expensive technique. However, the sludge volume is increased and further dewatering may be made more difficult because of the presence of coagulants. The difficulty with dewatering of sludges will obviously vary with the coagulants employed.

A great advantage of membrane systems is their ability to concentrate without the aid of chemical coagulants. Hence a system employing a membrane for thickening and a centrifuge for dewatering can produce a high-quality product free of other materials. Whether the capture of algae for reuse is important or not, the reduction in size of the dewatering equipment is a substantial asset.

2. Membrane Treatment for Potable Water Supplies

Low-pressure membrane systems may also be employed to produce high-quality waters from a variety of supplies for drinking purposes. To satisfactorily treat most surface supplies for potable water purposes, four objectives must be accomplished:

1. Color must be removed.
2. Turbidity and large-particulate materials must be removed.
3. Pathogenic bacteria and viruses must be eliminated or killed.
4. Objectionable tastes or odors must be eliminated.

Most color-causing contaminants in surface waters are negatively charged colloids that can be removed with open films. Turbidity is also caused by colloidal particles that can be easily removed with ultrafiltration films.

A carbon injection system can be employed to remove small molecules that create tastes and odors although a slip-stream treatment scheme is required for the treatment or disposal of spent carbon.

Dorr-Oliver workers[13] have reported on such a system indicating the production of a high-quality effluent. There was, however, a substantial problem in maintaining flux at satisfactory levels. Despite the fouling problem, there is an advantage of such treatment in that it does not produce a sludge that requires special disposal techniques. A system with high integrity should produce water that is substantially free of bacteria and viruses.

3. Sewage Treatment Plant Effluent Polishing

Hindin, Dunstan, and Bennett[14] made a comprehensive study of the reclamation of trickling filter and activated sludge effluent, as well as carrying out a number of single- and multicomponent studies on well-defined substances. The work is especially useful in that a variety of organics were studied both as single substrates and combinations of several organics. In addition, both bacteria and phage removal were examined. Cellulose acetate films with substantial sodium chloride retention were employed in the work.

In a general way, the results of the study agreed well with the earlier work reported by Sourirajan,[15] Loeb and Manjikian,[16] and the more recent studies of Davies.[17]

The reduction in ionic species in the effluent in this work varied from 80 to 90%. In each case, a water of high quality free from phage and, with one exception, free from coliform organisms was produced. Nitrate, as could be expected, was poorly retained but was held within acceptable limits.[18]

The work also reflects some interesting but rather inexplicable examples of synergism and antagonism in the retention of individual components of mixtures of organics. The lesson to be learned is that each specific mixture of organic wastes must be examined separately because each will have unique characteristics.

The data support the contention that separation of species similar in size can be made on the basis of pH effects. A marked difference in chlorophenol retention has been noted between treatment at pH 7.5 and at pH 9.0. A slight but consistent change in phenol retention was also noted. Phenolates appear to permeate less readily than the molecule with the proton intact. An acidic environment promotes transport of phenol.

The Federal Water Quality Administration has examined reverse osmosis as a part of their advanced treatment study program.[19] Their work was essentially a study of the removal of trace materials from effluents, similar in many ways to the work of Hindin et al.,[14] and as a consequence, was essentially a brackish water study. The results of the federal work were similar but less complete than the Hindin study.

Recently, a membrane study program has been undertaken at the University of Massachusetts.[20] The first phase of the study was to determine the removal of pesticides with reverse-osmosis membranes. In this regard and in the study of phenols, the work overlapped the study of Hindin. Important differences were observed in the work on lindane. At low pressures a marked decrease in the retention of lindane was noted with time. There was an apparent buildup of the pesticide which bleeds through the film.

The permeation of lindane through a film whose prime mode of transport is diffusion does not fit with the accepted theory of hydrogen-bond transport. Lindane certainly does not form hydrogen bonds to any extent, but substantial transport was observed at pressures of 30–50 psi.

The University of Massachusetts workers noted the substantial variation between their results on lindane and those of Hindin. In other cases where similar or identical compounds were studied, results were comparable.

V. AN ANALYSIS OF SYSTEM ECONOMICS

A. Factors Affecting Economics

An attempt to analyze membrane-system economics in the industrial waste treatment field in any depth is frustrated by the paucity of cost data available. Of equal importance is that many of the reported costs are based on studies in relatively small-scale equipment, and future costs almost certainly will be favorably affected by increases in production scale. This factor has been noted by other authors[11, 12] as well as by the manufacturers of membrane systems.

The principal factors to be considered in an economic analysis are the following:

1. The capital cost of the system: (*a*) the flux anticipated; (*b*) the complexity of the system; and (*c*) components required other than membranes.
2. Anticipated membrane life.
3. Membrane replacement costs.
4. Labor costs (including time for membrane replacement).
5. Power requirements.
6. Concentrate stream treatment or disposal costs.

B. Some Examples of Systems Costs

Membrane systems have some special and unusual aspects which must be considered in an analysis of system costs. These factors are as follows:

1. The generalization that capital costs are proportional to flow rate and operating costs to concentrations of impurities is not necessarily applicable to membrane systems.

2. Because of the modular character of membrane plants, reductions in cost normally associated with an increase in throughput are much smaller than are usual for the process industries.

3. No manufacturer is currently producing membranes or modules on a scale that will permit cost minimization. Hence, present costs are high by an unpredictable (but very real) multiplier.

Table 7. Capital and Operating Costs for Membrane Sewage Treatment Works

Capacity (gal/day)	Capital Cost ($/kgal) Complete Treatment	Tertiary Treatment	Operating Cost ($/kgal) Complete Treatment	Tertiary Treatment
3,600	3,300		1.61	
10,000	1,900	1,400	1.40	
20,000	1,500	1,200		0.69
30,000	1,370	1,100	1.23[a]	
40,000	1,300	1,070		0.66[a]
50,000	1,240	1,060		
60,000	1,200	1,050	1.16[a]	0.61[a]
80,000	1,160[a]	1,040[a]		
100,000	1,120[a]	1,030[a]	1.00[a]	0.55[a]

[a] Computed from operating and estimating price data.

The lack of scale effect is noted in the report of Weissman, Smith, and Okey,[13] wherein the capital costs of membrane sewage treatment facilities are presented. These costs are for the Dorr-Oliver type of unit employing ultrafiltration films. The cost data are presented in Table 7. The costs for conventional waste-water treatment plants would decrease almost an order of magnitude across the flow range covered in Table 7. The small modular character of the membrane system results in a much smaller reduction in costs.

The membranes employed in the systems for which costs are given in Table 7 have demonstrated a mean life of more than 6 months. The cutoff range* was between 10,000 and 30,000, the system pressure was between 15 and 40 psi; the flux was between 30 and 8 gal/(day)(ft²). Significantly, the flux with clear water for these films was greater than 200 gal/(day)(ft²) at 50 psi.

Rickles[21] reported that a 1,000,000 gal/day system with desalination capability could be owned and operated for $0.40/kgal. The cost for a 10,000,000 gal/day facility was estimated at $0.30/kgal. The amortized capital costs varied between $0.08 and $0.15/kgal for the two plants. These costs for amortization are based on long-term amortization schedules— generally greater than 20 yr.

Recently, with plants having the capability to remove in excess of 99% of divalent metal ions, the costs were found to be between $0.90 and $3.00/ kgal, depending on size and character of the metals removed. These costs were for plants under 1,000,000 gal/day and generally included a subsystem for treating the concentrate.[22] A major part of the cost is membrane replacement. It is expected that membrane life and replacement costs will be altered to reduce the effective overall cost to less than 50% of the costs determined here. Some companies are now guaranteeing 5-yr life at a specific flux. This has done a great deal to promote interest in membrane systems. Membrane life was taken at 2 yr, a 10-yr amortization schedule was employed and fluxes were assumed to vary between 11 and 7 gal/(ft²)(day) at 25°C across the life of the films.

The polishing of sewage effluents for reuse is essentially a desalination application. The costs of such operations will be similar to the costs for brackish water treatment. The costs are similar to the costs for tertiary treatment, desalination, and water softening. In general, to own and operate such a facility costs between $0.40 and $1.00/kgal in sizes from 0.5 to 1.0 × 10⁶ gal/day. Potable water produced with a membrane system from moderately contaminated surface waters should cost between $1.00 and $3.00/kgal, depending upon system size. The cost of conventional water treatment will vary between $0.10 and $0.50/kgal without special provision for sludge treatment. These costs may be doubled if special sludge treatment measures are required. Membrane systems have an added advantage in that a genuine logistic advantage is realized by the system since chemicals are not needed. The special applications for such systems are extensive.

In general, to obtain the price information presented in the foregoing paragraphs, the following values were assigned the pertinent parameters:

1. Membrane life—1–2 yr.
2. Membrane flux—10–20 gal/(ft²)(day).
3. Amortization—5–10 yr for industrial applications and more than 20 yr for municipal applications.

* The range of molecular weights that would be retained.

C. Some General Conclusions Regarding Economics

Depending on the amortization schedule, but assuming that it is five years at currently accepted interest rates, some important conclusions can be drawn from the cost data presented in this section:

1. Membrane replacement is the most important individual cost item with the present state of the art. This item appears to vary from $0.40 to $2.00/kgal.

2. Capital cost is a significant but not a controlling cost factor. The amortized capital costs for waste treatment facilities appeared to vary from $0.15 to $1.00/kgal.

3. The estimated capital costs of membrane systems were found to vary as much as 100% from one manufacturer to another. This factor apparently relates to the fact that membrane technology is only in its infancy, so to speak.

4. Based on the cost data presented in this chapter, there are three categories of applicability of membrane processes. (a) Membrane treatment is clearly not applicable for treatment of wastes in which there are large flows of dilute, easily bioassimilable waste streams from which little or no value can be recovered to offset costs. Examples are vegetable and fruit processing wastes, dilute chemical wastes, irrigation return flows, and acid mine drainage. (b) Membrane treatment is clearly indicated for wastes that meet one or more of the following qualifications: high-value waste components, high concentrations, low-to-moderate flows, need for space conservation, and situations in which product recovery is feasible on the concentrate stream. Examples are treatment of plating wastes, treatment of some chemical wastes that primarily contain monomers or dimers suitable for reuse, and treatment of some pharmaceutical wastes. (c) There are marginal cases in which special logistic problems are encountered, sludge disposal may be a problem, or space and configuration requirements mitigate against conventional treatment. Examples are water treatment, water softening, special waste treatment package system (domestic and industrial), and special single-component waste streams.

VI. SUMMARY

Pressure-driven membrane systems offer a waste-management tool of unlimited potential. Whether the potential is to be realized depends on the development of systems that can maintain a substantial fraction of the flux obtained with clear-water feeds, when the systems are treating wastes.

With the present state of development there are several applications that can profitably employ pressure-driven membrane systems as a waste-treatment device. Some of these applications are for treatment of wastes in which:

1. Gravity separation cannot be employed.
2. A high-value product is normally lost in the waste stream.
3. Space or configurational considerations prohibit conventional schemes.
4. There is need or value in excluding or passing a specific waste component.
5. Mobility of the treatment system is important.

Treatment costs vary from $0.30 to $0.40/kgal for the simplest systems to several dollars per thousand gallons for complex high-pressure systems.
Broad general acceptance must await further system improvements.

REFERENCES

1. *The Cost of Clean Water and Its Economic Impact*, Vol. IV, *Projected Wastewater Treatment Costs in the Organic Chemicals Industry*, Federal Water Quality Administration; U.S. Dept. of the Interior, Washington, D.C., June 1968.
2. W. W. Eckenfelder, Jr., *Effluent Quality and Treatment Economics for Industrial Wastewaters*, Federal Water Quality Administration, U.S. Dept. of the Interior, Washington, D.C., October 1967.
3. *The Cost of Clean Water and Its Economic Impact*, Vol. 1, Federal Water Quality Administration, U.S. Dept. of the Interior, Washington, D.C., January 1969.
4. *Waste Treatment Processes*, Dorr-Oliver, Inc., Stamford, Conn., 1966, Charts 1, 2, and 3.
5. R. W. Okey, "The Application of Membranes to Sewage and Waste Treatment," paper presented at University of Texas Symposium on Physical and Chemical Processes in Waste Treatment, Austin, Texas, March 1969.
6. R. A. Fiedler and R. W. Okey, "An Evaluation of Diffusion Membranes for Waste Water Rehabilitation," Texas Industrial Wastes Conference, University of Texas, Austin, Texas, 1967.
7. A. C. F. Ammerlaan, B. F. Lueck, and A. J. Wiley, "Membrane Processing of Dilute Pulping Wastes by Reverse Osmosis," paper presented at TAPPI Water Conference, Portland, Oregon, April 1968.
8. A. J. Wiley, A. C. F. Ammerlaan, and G. A. Dubey, "Application of Reverse Osmosis to Processing of Spent Liquors from the Pulp and Paper Industry," *Tappi* **50** (9) (September 1967).
9. R. W. Okey and P. L. Stavenger, "Reverse Osmosis Application in Industrial Waste Treatment," *Membrane Processes for Industry*, Southern Research Institute, Birmingham, Ala., 1966, pp. 127–156.
10. H. J. Bixler, R. W. Hausslein, and L. Nelsen, "Separation and Purification of Biological Materials by Ultrafiltration," paper presented at the 65th National American Institute of Chemical Engineers Meeting, Cleveland, Ohio, May 4–7, 1969.
11. R. W. Okey, P. L. Stavenger, and D. S. Davies, "Engineered Membrane Systems, A New Dimension in Waste Management," paper presented at American Chemical Society, Miami Beach, Florida, April, 1967.

12. R. L. Goldsmith, "Macromolecular Ultrafiltration with Microporous Membranes," ABCOR, Inc., Cambridge, Mass., 1969.

13. B. J. Weissman, C. V. Smith, Jr., and R. W. Okey, "Performance of Membrane Systems in Treating Water and Sewage," paper presented at American Institute of Chemical Engineers Meeting, Los Angeles, California, December 1968.

14. E. Hindin, G. H. Dunstan, and P. J. Bennett, "Water Reclamation by Reverse Osmosis," Federal Water Quality Administration, U.S. Dept. of the Interior, Washington, D.C., August 1968.

15. S. Sourirajan, "Reverse Osmosis Separation and Concentration of Sucrose in Aqueous Solutions Using Porous Cellulose Acetate Membranes," *Ind. Eng. Chem. Proc. Des. Dev.* **6** (1) (1967).

16. S. Loeb and S. Manjikian, "Field Tests on Osmotic Desalination Membranes," *UCLA, Dept. Eng. Rep. 64-34,* 1964.

17. D. S. Davies, "Transport Properties of Reverse Osmosis Membranes for Mixed Organic-Inorganic Aqueous Solutions," M.S. Thesis, University of Connecticut, 1968.

18. T. S. Govendan and S. Sourirajan, "Separation of Some Inorganic Salts in Aqueous Solution Using Porous Cellulose Acetate Membranes," *Ind. Eng. Chem. Proc. Des. Dev.* **5** (1966).

19. "The Advanced Waste Treatment Research Program, January 1962 through June 1964," *U.S. Public Health Serv. Publ. No. 999-WP-24,* April 1965.

20. R. T. Skrinde, T. H. Feng, P. A. Lutin, and W. L. Short, "Low Pressure Ultra-filtration Systems for Wastewater Contaminant Removal," Federal Water Quality Administration, U.S. Dept. of the Interior, Washington, D.C., March 1969.

21. R. N. Rickles, *Membrane Technology*, Noyes Development, Park Ridge, N.J., 1967.

22. Resource Engineering Associates, Wilton, Connecticut, October 1969, unpublished data.

Chapter XIII   Gas Permeation Processes

S. A. Stern*

I. INTRODUCTION

The separation of gases and vapors by selective permeation through non-porous membranes is a technological concept well over one hundred years old, which has remained curiously neglected through most of its long history. The first observation that different gases permeate through membranes at unequal rates is generally attributed to Mitchell,[1] who reported this phenomenon in 1831. Thirty-five years later, Graham[2] discussed the mechanism of permeation in strikingly modern terms and demonstrated experimentally that mixtures of gases can be separated by means of rubber membranes. Following Graham's work, selective permeation was considered from time to time as a practical gas separation technique, and at least two patents on air separation were issued before the end of the last century.[3, 4] The permeation of gases, particularly through rubber, was also studied during the first quarter of this century in relation to other uses, such as the leakage of air from automobile tires and of helium and hydrogen from the envelopes of lighter-than-air craft.

Despite this early work, no serious attempt to appraise the potential of gas permeation as a large-scale separation technique was made until after World War II. At that time, many new types of polymeric membranes were developed and manufactured in sizable quantities for packaging purposes and other commercial applications. The availability of these materials stimulated renewed interest in their usefulness as separation barriers. Thus in the early 1950's, Weller examined for the first time the engineering aspects of

* Department of Chemical Engineering and Materials Science Syracuse University, Syracuse, New York.

several gas permeation processes of industrial importance, namely, the separation of oxygen from air, the recovery of helium from natural gas, and the separation of hydrogen from coal-hydrogenation tail gas and refinery gas.[5, 6] This work was extended by Kammermeyer and his associates to other separation processes, such as the concentration of ammonia from mixtures with nitrogen and hydrogen, and the separation of carbon dioxide from various gas mixtures.[7-9]

Since the pioneering studies of Weller and Kammermeyer, considerable progress has been made in all facets of gas permeation technology. For example, new membranes exhibiting both higher permeabilities and selectivities have been synthesized, special equipment for large-scale separations has been developed and tested, and experience has been gained in process design and optimization. Although this work has been concerned primarily with organic polymeric membranes, inorganic and metallic types of membranes and barriers have also been studied. As a result, the factors determining the economics of selective permeation are much better understood and more realistic evaluations of this separation method are now possible.

This chapter attempts to review the state-of-the-art of permeation technology, insofar as it applies to the separation of gases and vapors by means of polymeric membranes. These membranes are considerably more versatile and, in the opinion of the writer, offer greater economic promise than other forms of separation barriers. The review is not comprehensive, nor can it be so, since much of the work in this field is proprietary and has not been reported in the open literature. Even some of the published material is of limited value, because it is either of a promotional nature or cannot be interpreted without further data. This is particularly true of economic studies and evaluations of specific processes. In view of the competitive pressures motivating the development of new separation techniques, a reluctance to reveal privileged information is understandable.

Finally, no claim is made that a well-balanced presentation of the subject has been achieved. The review necessarily reflects the interests and prejudices of its writer.

II. THE RATE OF GAS PERMEATION

A.   General Considerations

The present knowledge about the mechanism of gas permeation through polymeric membranes has been surveyed in varying detail by van Amerongen,[10] by Rogers,[11] and in a monograph edited by Crank and Park.[12] These surveys were made from the viewpoints of the physical chemist or the polymer scientist and do not require further elaboration. However, it is useful to re-examine certain aspects of the permeation mechanism of importance to the designer of separation processes. For example, the designer needs a description of the effect of pressure and temperature on the rate of gas permeation over wide ranges of these variables. He also requires detailed information on the permeation of gas mixtures under conditions of practical interest. Unfortunately, such information is scarce and the discussion of these subjects must rest largely on phenomenological considerations.

The permeation of a gas through a nonporous polymeric membrane is a complex process that may involve the following sequence of steps: (a) adsorption of the gas at one interface of the membrane, (b) solution of the gas into the membrane at that interface, (c) activated diffusion of the gas in and through the membrane, (d) release of the gas from solution at the opposite interface, and finally, (e) desorption from the latter interface. Steps a and b, as well as d and e, are not necessarily distinct. The term *permeation* is therefore used to describe the overall mass transport of gas across the membrane, whereas the term *diffusion* refers only to the movement of the gas inside the membrane matrix.

Two other general statements can be made concerning the permeation of gases through polymeric membranes: gases permeate in molecular form (unless they are monatomic), and all membranes are permeable to various extents to all gases.

Barrer[13] and other investigators showed that activated diffusion is usually the rate-controlling step in the permeation process. Diffusion of a gas in a membrane can be described by Fick's first law, which takes the following form for one-dimensional transport in a direction normal to the membrane interfaces:

$$J = -D\left(\frac{\partial c}{\partial x}\right) \tag{1}$$

where J is the rate of diffusion of the penetrant gas through a unit reference area; D is the diffusion coefficient for a specific penetrant–membrane system and temperature; and c is the concentration of the penetrant in the membrane at a position coordinate x. The above relation is generally valid only for isotropic media.

According to Equation 1, the flux J is proportional to both the diffusion coefficient and the concentration gradient, $\partial c/\partial x$, measured normal to the reference area. The diffusion coefficient for systems of gases and polymeric membranes can be constant or a function of penetrant concentration. It can also depend on position coordinate, on time, and on previous "history"; in the last case the diffusion is called "anomalous" or "non-Fickian." The concentration gradient can be obtained from Fick's second law, which is given for one-dimensional diffusion by

$$\frac{\partial c}{\partial t} = \frac{d}{\partial x}\left(D\frac{\partial c}{\partial x}\right) \qquad (2)$$

where $\partial c/\partial t$ is the rate of change in concentration with time, t, at a position coordinate x. Solutions of this partial differential equation have been summarized by Barrer,[13] Crank,[14] and Jost[15] for constant as well as variable diffusion coefficients, for a variety of geometries and boundary conditions, and for both steady-state and transient flow. Experimental methods for the determination of diffusion coefficients have been reviewed by the above authors and by others.

The concentration of a gaseous penetrant in a polymeric membrane is dependent on the solubility of the penetrant in the polymer. Under conditions of solution equilibrium, the relation between the pressure p of the penetrant in the gas phase and its (uniform) concentration c in the polymer is usually expressed in the form

$$c = S(c)p \qquad (3)$$

or

$$c = S(p)p \qquad (4)$$

where $S(c)$ and $S(p)$ are solubility coefficients reported as functions of concentration or pressure. When the penetrant solubility is sufficiently low, Equations 3 and 4 reduce to Henry's law

$$c = S_0 p \qquad (5)$$

where the solubility coefficient S_0 is a constant for any specific penetrant–membrane system and temperature; as defined here, S_0 is the reciprocal of the conventional Henry's law constant. Relations 3, 4, and 5 are measured under isothermal conditions and are commonly called *absorption isotherms*.

Most gas–membrane systems can be classified into four major categories, depending on the properties of the diffusion and solubility coefficients. According to Barrer,[16] these categories are described by the following sets of relations:

(I) $\qquad \dfrac{\partial c}{\partial t} = D\dfrac{\partial^2 c}{\partial x^2}$; $\qquad D = D_0$; $\qquad c = S_0 p$

(Henry's law obeyed)

(II) $\qquad \dfrac{\partial c}{\partial t} = \dfrac{\partial}{\partial x}\left(D\dfrac{\partial c}{\partial x}\right)$; $\qquad D = D(c)$; $\qquad c = S_0 p$

(Henry's law obeyed)

(III) $$\frac{\partial c}{\partial t} = \frac{\partial}{\partial x}\left(D\,\frac{\partial c}{\partial x}\right); \quad D = D(c); \quad c = S(c)p$$

(Henry's law not obeyed)

(IV) $$\frac{\partial c}{\partial t} = \frac{\partial}{\partial x}\left(D\,\frac{\partial c}{\partial x}\right); \quad D = D(c,t); \quad c = S(c)p$$

(Henry's law not obeyed)

D_0, $D(c)$, and $D(c,t)$ denote, respectively, that the diffusion coefficient is a constant, a function of concentration, or a function of both concentration and time.

The first two categories comprise systems of penetrants and polymeric membranes of particular interest in gas separation processes. Examples of systems falling in all categories are given below. However, these examples must be considered as only tentative because of the scarcity of diffusion and solubility measurements over wider ranges of concentration (or pressure) and temperature. Most of the pertinent data were obtained at temperatures generally not more than about 30°C above or below the ambient, and the classification pertains to this temperature range. Considerable overlapping must be anticipated in such a classification.

Category I represents diffusion and solution of simple gases in elastomers and many of the harder polymers at pressures up to several atmospheres, and in some cases up to much higher pressures.[12, 16] These gases are characterized by critical temperatures that are low compared to the ambient temperature, as is the case for the so-called "permanent" gases. Both the diffusion and the solubility coefficients for this category are constants.

Category II includes systems of penetrants with higher critical temperatures, from near ambient to perhaps as high as 200°C, in both elastomers and harder polymers. The diffusion coefficient for such systems is a function of the penetrant concentration in the membrane, but the solubility coefficient is constant; that is, Henry's law is still obeyed. This category is exemplified by the diffusion and solution of C_4 and C_5 paraffins in rubber membranes.[17] Other examples are given in the following section.

For penetrant–membrane systems in Category III, both the diffusion and the solubility coefficients are functions of concentration. The penetrants in this category are gases with high critical temperatures, usually above 200°C. Examples can be found in systems of many organic vapors, such as benzene, chloroform, p-xylene, C_5–C_8 paraffins, and methyl bromide in polyethylene,[18] or of heavier hydrocarbon vapors and chloroform in rubbers with high chain mobility and short relaxation times.[19–23]

Lastly, systems in Category IV are characterized by time- and history-dependent diffusion, as has been observed with various organic vapors in ethyl cellulose.[24, 25] Polymers in this category are more rigid and internally viscous, and exhibit longer relaxation times than those in previous categories. Time- and history-dependent transport behavior has also been observed for systems in the first category below the glass-transition temperature of the polymer.[26]

The above classification is clearly not inclusive, and further categories of penetrant–membrane systems can be recognized.[16] Thus an additional category could represent systems for which the diffusion coefficient is a function of position in the membrane. Such behavior could be expected for composite membranes consisting of two or more layers for which the diffusion coefficients are different. This category may prove to be of considerable importance in gas separation processes.

B.   Effect of Pressure

The rate of mass transport of penetrant *across* a membrane, that is, the rate of permeation, is most easily determined for penetrant–membrane systems of Category I. Permeation through planar membranes and under steady-state conditions is perhaps of greatest practical interest. These conditions can be achieved by maintaining the concentrations of the penetrant at the two interfaces of the membrane at different but constant values. The rate of permeation is then also constant and, for a diffusion-controlled process, can be expressed by the following form of Equation 1:

$$J = -D_0 \left(\frac{dc}{dx}\right) = \text{constant} \tag{6}$$

From this relation it follows that

$$\frac{dc}{dx} = \text{constant} \tag{7}$$

which shows that the penetrant concentration decreases linearly across the membrane, as illustrated in Figure 1.

Equation 6 can be integrated across the thickness of the membrane, δ, to yield

$$J = D_0 \frac{(c_h - c_l)}{\delta} \tag{8}$$

where c_h and c_l are the penetrant concentrations at the membrane interfaces; the subscripts h and l signify "high" and "low" ($c_h > c_l$). It is more practical to express the rate of permeation in terms of the penetrant gas pressures p_h and p_l on the two sides of the membrane, in equilibrium with c_h and c_l, respectively. Thus substitution of Equation 5 in Equation 8 results in the expression

$$J = (D_0 S_0) \frac{(p_h - p_l)}{\delta} \tag{9}$$

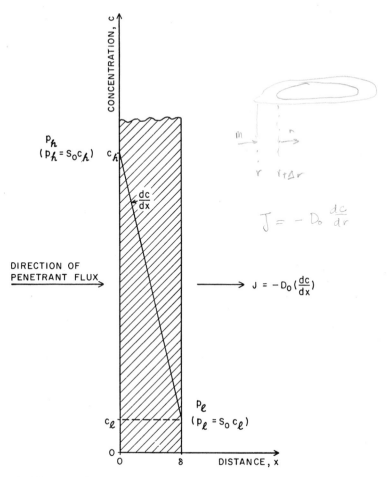

Figure 1. Concentration gradient across membrane for constant diffusion and solubility coefficients. J, Penetrant flux; p, pressure; c, concentration; δ, membrane thickness; $D_0 =$ diffusion coefficient; S, solubility coefficient. Subscripts: h, high-pressure side of membrane; l, low-pressure side of membrane. D_0 and P_0 are constants.

Finally, the total rate of gas permeation, G, through a planar membrane of area A is derived for steady-state conditions from Equation 9:

$$G = JA = P_0 \frac{A\Delta p}{\delta} \tag{10}$$

where

$$P_0 \equiv D_0 S_0 \tag{11}$$

P_0 is usually called the "permeability coefficient," the "permeability constant," or simply the "permeability," and $\Delta p (= p_h - p_l)$ is the difference between the pressures on the two sides of the membrane. The permeability coefficient depends only on the nature of penetrant–membrane system and the temperature. Consequently, the rate of permeation through a planar membrane is directly proportional to the permeability coefficient, the membrane area, and the pressure differential, and is inversely proportional to the membrane thickness.

Permeation through a tubular membrane is another case of practical interest. If the inner and outer radii of the tube are r_I and r_O, respectively, the rate expression corresponding to Equation 10 is

$$G = P_0 \frac{2\pi h (p_h - p_l)}{\ln (r_O/r_I)} \tag{12}$$

where h is the length of the tube. In this case, the concentration of the penetrant in the membrane does not vary linearly with distance.

For penetrant–membrane systems in other categories, Fick's equations can be integrated only if the dependence of the diffusion coefficient on concentration is known. If this information is not available, the rate of permeation must be expressed in terms of a mean, or integral, diffusion coefficient, $\bar{D}$. For instance, in the case of a concentration-dependent diffusion coefficient, integration of Equation 1 for a planar membrane yields for steady-state conditions:

$$J = \frac{1}{\delta} \int_{c_l}^{c_h} D(c)\, dc = \bar{D} \frac{(c_h - c_l)}{\delta} = \text{constant} \tag{13}$$

where

$$\bar{D} = \frac{1}{(c_h - c_l)} \int_{c_l}^{c_h} D(c)\, dc \tag{14}$$

According to Rogers,[11] the rate of permeation can then be written as a function of the pressure difference between the two sides of the membrane by multiplying and dividing Equation 13 by $(p_h - p_l)$,

$$J = \bar{D} \left(\frac{c_h - c_l}{p_h - p_l} \right) \left(\frac{p_h - p_l}{\delta} \right) = (\bar{D}\bar{S}) \frac{(p_h - p_l)}{\delta} \tag{15}$$

where

$$\bar{S} = \frac{c_h - c_l}{p_h - p_l} \tag{16}$$

$\bar{S}$ is a quantity related to the solubility of the penetrant in the membrane.

The overall rate of permeation, G, through a planar membrane of area A is obtained from Equation 15:

$$G = JA = \bar{P}\frac{A(p_h - p_l)}{\delta} \tag{17}$$

where

$$\bar{P} \equiv \bar{D}\bar{S} \tag{18}$$

$\bar{P}$ is a mean or integral permeability coefficient. For any penetrant–membrane system, $\bar{P}$ is a function of the pressure on the two interfaces of the membrane, as well as of the temperature. Information on the explicit forms of this functionality is very limited, but can be surmised from the following considerations.

Many permeation studies have been conducted under conditions where the low-pressure side of the membrane is maintained at near-zero pressure, such that $p_h \gg p_l \approx 0$ and $c_h \gg c_l \approx 0$. The mean diffusion coefficient is then defined by

$$\bar{D} = \frac{1}{c}\int_0^c D(c)\,dc \tag{19}$$

and the quantity $\bar{S}$ becomes

$$\bar{S} = \frac{c}{p} = S \tag{20}$$

where $c = c_h$ and $p = p_h$; hence $\bar{S}$ reduces to the solubility coefficient of the penetrant in the polymer, S, in accordance with Equations 3 or 4. Rogers, Stannett, and Szwarc[18] have found experimentally for many organic vapors in polyethylene that in this case $\bar{D}$ can be expressed as a function of pressure by the empirical relation

$$\bar{D} = D(0)\exp(\alpha a) = D(0)\exp\left[\alpha\left(\frac{p}{p^*}\right)\right] \tag{21}$$

where a is the vapor activity, α is a constant, and $D(0)$ is the diffusion coefficient at zero activity, pressure, and concentration. The activity is given approximately by

$$a \simeq \frac{p}{p^*} \tag{22}$$

where p^* is the vapor pressure of the pure penetrant. These investigators also found from equilibrium sorption measurements that

$$S = \frac{c}{p} = S(0)\exp(\sigma c) \tag{23}$$

where $S(0)$ is the solubility coefficient at zero concentration or pressure, and σ is a constant that defines the concentration dependence of S. The relation above implies that Henry's law is obeyed only in the limit $c \to 0$. The solubility data of Rogers, Stannett, and Szwarc[27] represented by Equation 23 can be also expressed satisfactorily as a function of penetrant pressure by the relation

$$S = S(0) \exp \left[\frac{\sigma S(0)p}{1 - \sigma S(0)p} \right] \quad (24)$$

Substituting Equations 20, 21, and 24 into Equation 18 finally yields

$$\bar{P} = P(0) \exp \left\{ \frac{\left[\frac{\alpha}{p^*} + S(0)\sigma \right] p - \left[\frac{\alpha S(0)\sigma}{p^*} \right] p^2}{1 - S(0)\sigma p} \right\} \quad (25)$$

where

$$P(0) \equiv D(0)S(0) \quad (26)$$

and where $P(0)$ is the permeability coefficient at zero pressure and concentration. $P(0)$, $D(0)$, and $S(0)$ are constants that depend only on the nature of the penetrant–membrane system and on the temperature. Equation 25 is an empirical representation of the permeability behavior of many penetrant–membrane systems in Category III. In accordance with Equation 25, plots of log $\bar{P}$ versus (p/p^*) are nonlinear for such systems, as shown in Figure 2 for several organic vapors in polyethylene.

For systems in Category II, Henry's law is obeyed ($\sigma = 0$) and Equation 25 reduces to

$$\bar{P} = P(0) \exp (\beta p) \quad (27)$$

where

$$\beta = \frac{\alpha}{p^*} \quad (28)$$

Hence, in this case, a plot of log $\bar{P}$ versus pressure or vapor activity should be linear at constant temperature. This has been observed, for example, for ethylene oxide in polyethylene and rubber hydrochloride at 0°C, as well as for methyl bromide in polytrifluorochoroethylene at 60°C and in rubber hydrochloride at 0°C.[28] It has also been observed over wide ranges of pressure for carbon dioxide, ethane, ethylene, and nitrous oxide in polyethylene,[27, 29, 30] but it is not known to what extent Henry's law is obeyed by these systems. At low pressures or vapor activities, Equation 27 reduces, in turn, to the linear form:

$$\bar{P} = P(0) + Kp \quad (29)$$

where $K = P(0)\alpha/p^*$; such a relationship has been reported by Cutler et al.[31] Equations 27 and 29 also describe the behavior of systems in Category III at sufficiently low vapor activities (or concentrations).

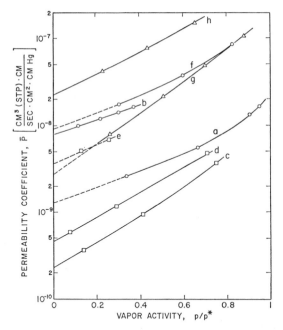

Figure 2. Permeability coefficients for various organic vapors in polyethylene; dependence of permeability coefficients on vapor activity.[18] *a*, Methyl bromide (0°C); *b*, methyl bromide (30°C); *c*, isobutylene (−8°C); *d*, isobutylene (0°C); *e*, isobutylene (30°C); *f*, benzene (0°C); *g*, *n*-hexane (0°C); *h*, *n*-hexane (30°C). p = pressure of penetrant; p^* = vapor pressure of penetrant at indicated temperature.

When both α and σ are zero, Equation 25 reduces to

$$\bar{P} = P(0) = P_0 \qquad (30)$$

where the permeability coefficient is independent of pressure at any given temperature, as expected for systems of Category I. Such behavior has been reported for many penetrant–membrane systems, but in most cases the measurements were made over very limited ranges of pressure. Stern et al.[27] found that the permeability coefficient for methane in polyethylene was independent of pressure up to 66 atm and at temperatures between 10 and 40°C, as illustrated in Figure 3; similarly, the permeability coefficient for carbon dioxide in polyethylene was reported to be essentially constant up to 54 atm, but only above the critical temperature of the penetrant.[29] However, it was found later that such results are uncommon and depend on the cancellation of opposing physical effects.[30]

Stern et al.[27, 30] also observed that the permeability of certain systems in Category I, such as helium and nitrogen in polyethylene, *decreased* slightly

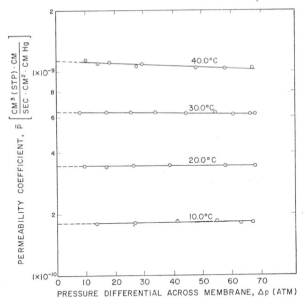

Figure 3. Permeability coefficients for methane in polyethylene—dependence of permeability coefficients on pressure differential across membrane.[28]

with increasing penetrant pressure. This rather unexpected observation cannot be understood on the basis of the previous considerations, but has been explained recently in terms of an extension of well-known "free volume" theory of diffusion in polymers.[30]

Although Equation 25 describes the effect of pressure on the permeability coefficients for many penetrant–membrane systems (when $p_h \gg p_l \approx 0$), it is limited by the particular dependence of $\bar{D}$ and S on pressure, or concentration, assumed in Equations 21 and 23. Thus the latter relation is based on absorption isotherms characterized by negative deviations from Henry's law. However, other types of isotherms have also been observed. For instance, the solubility of nitrogen and methane in polyethylene was found to exhibit positive deviations from Henry's law at 25°C, that is, the solubility coefficient decreased with increasing pressure, when measured over a sufficiently wide pressure range (up to 80 atm).[32] The solubility of hydrogen sulfide in cellulose acetate at 0°C and in ethyl cellulose at −35 and 0°C exhibits a similar behavior.[33] On the other hand, the absorption isotherm for water vapor in hydrophilic polyamide membranes at 25°C is S-shaped, probably due to the existence of strong hydrogen-bonding sites in the polymer.[34]

It is interesting to note another method of describing the effect of pressure on gas permeation. Equations 21 and 23 can be combined into

$$\bar{D} = D(0) \exp \left[\frac{\gamma c}{\exp (\sigma c)} \right] \qquad (31)$$

where

$$\gamma = \frac{\alpha}{p^* S(0)} \qquad (32)$$

If Henry's law is obeyed, Equation 31 reduces to

$$\bar{D} = D(0) \exp (\gamma c) \qquad (33)$$

which has been used by many investigators. Moreover, differentiation of Equation 19 with respect to concentration and substitution of Equation 33 for $\bar{D}$ yields[18]:

$$D = D(0)(1 + \gamma c) \exp (\alpha c) \simeq D(0) \exp (bc) \qquad (34)$$

where b is a constant. This exponential dependence of the *differential* diffusion coefficient, D, on concentration has also been reported for many penetrant–membrane systems. If Equation 34 is substituted for $D(c)$ in Equation 13, the latter can be integrated between the concentration limits c_h and c_l at the membrane interfaces. The following relation is then obtained for the penetrant flux:

$$J = \frac{1}{\delta} \int_{c_l}^{c_h} D(0) \exp (bc) \, dc = \frac{D(0)}{b\delta} [\exp (bc_h) - \exp (bc_l)] \qquad (35)$$

Since the solubility of the penetrant in the polymer was assumed to obey Henry's law, the total rate of permeation G across membrane area A can be derived from Equations 5 and 35 in terms of the pressures p_h and p_l on the two sides of the membrane:

$$G = JA = \frac{D(0)A}{b\delta} [\exp (kp_h) - \exp (kp_l)] \qquad (36)$$

where $k = bS_0$. Li and Long measured the permeabilities of methane and ethylene in polyethylene at elevated pressures and correlated their data by a similar relation.[32]

C. Effect of Temperature

Both the diffusion and the solubility coefficients for penetrant–membrane systems of Category I are usually exponential functions of temperature and can be expressed by the following Arrhenius-type relations:

$$D_0 = D_0' \exp \left(\frac{-E_d}{RT} \right) \tag{37}$$

and

$$S_0 = S_0' \exp \left(\frac{-\Delta H_s}{RT} \right) \tag{38}$$

where E_d is the energy of activation for diffusion, ΔH_s is the enthalpy of solution, R is the universal gas constant, T is the absolute temperature, and D_0' and S_0' are constants. By virtue of Equation 11, the permeability coefficient is also an exponential function of temperature.

$$P_0 = P_0' \exp \left(\frac{-E_p}{RT} \right) \tag{39}$$

where E_p is the energy of activation of the permeation process and P_0' is a constant.

From Equations 11, 37, 38, and 39, it is seen that

$$P_0' = D_0' \cdot S_0' \tag{40}$$

and

$$E_p = E_d + \Delta H_s \tag{41}$$

The sign of E_p in Equation 41 depends on the following factors. The solution of a gas in a polymer can be visualized as a two-step process involving (a) condensation of the gas, which occurs exothermally, followed by (b) mixing of the condensate with the polymer, which is an endothermal step.[11, 35, 36] Hence, ΔH_s can be considered as the sum of two terms:

$$\Delta H_s = \Delta H_{cond} + \overline{\Delta H_1} \tag{42}$$

where ΔH_{cond} is the enthalpy of condensation of the penetrant gas, and $\overline{\Delta H_1}$ is the partial molal enthalpy of mixing. For simple gases above their critical temperature, ΔH_{cond} is a hypothetical quantity and its value is relatively small; hence ΔH_s will be approximately equal to $\overline{\Delta H_1}$ and positive. Since E_d is always a positive quantity, E_p will also be positive and, according to Equation 39, the permeability coefficient for penetrant–membrane systems in Category I will increase exponentially when the temperature is raised. This behavior is shown in Figure 4 where Equation 39 is plotted in the form of log P_0 versus $1/T$ for helium, hydrogen, nitrogen, oxygen, and carbon dioxide in polyethylene between 0 and 50°C.[37] Many similar examples have been reported in the literature. Over wider ranges of temperature, the plots of log P_0 versus $1/T$ are sometimes nonlinear, as has been reported by van

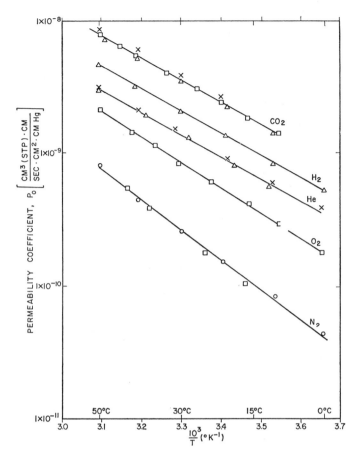

Figure 4. Permeability coefficients for various gases in polyethylene—dependence of permeability coefficients on temperature.[27]

Amerongen for helium, hydrogen, and nitrogen in natural rubber and Hycar OR 15 (a copolymer of 61 % butadiene and 39 % acrylonitrile); in this case, the plots are concave with respect to the $1/T$ coordinate.[36]

Distinct changes in the slope of the Arrhenius plots for both the diffusion and permeability coefficients have been observed at temperatures at which the polymer undergoes second-order transitions, such as the "glass" transition. Meares has measured these coefficients for helium, hydrogen, neon, oxygen, and krypton in polyvinyl acetate[38, 39] and found that D was independent of pressure ($D = D_0$) and Henry's law was obeyed ($S = S_0$) for the entire range of his experimental conditions. However, the plots of log P_0

versus $1/T$ and log D_0 versus $1/T$ for each penetrant could be represented by three distinct straight lines with two intersections or "breaks" at 26° and 18–15°C, corresponding to two transitions in polyvinyl acetate. The lower transition temperature tended to decrease as the molecular size of the penetrant increased.

With partially crystalline membranes, anomalous or non-Fickian behavior is sometimes observed below the glass-transition temperature of the polymer.

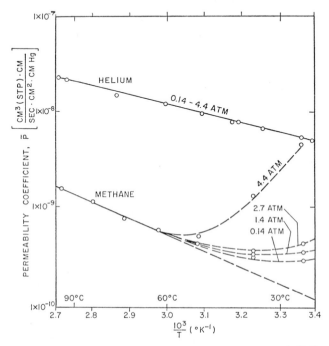

Figure 5. Permeability coefficients for helium and methane in Teflon FEP. Effect of glass-transition temperature.[26]

This is illustrated in Figure 5 for the permeation of methane through Teflon FEP, a copolymer of tetrafluoroethylene and hexafluoropropylene.[26] At temperatures higher than about 60°C, the permeability coefficient was found to be independent of pressure at least up to 50 atm ($\bar{P} = P_0$). Below this temperature, the permeability coefficient was dependent on pressure, time, and permeation "history." For the example shown in Figure 5, the high values of $\bar{P}$ obtained for methane at 4.4 atm and below 60°C were found to decrease with time and approach the values expected from a linear extrapolation of the log $\bar{P}$ versus $1/T$ plot above 60°C. Similar behavior, which is probably a result of relaxation phenomena in the polymer, was observed with nitrogen in Teflon FEP.[26] On the other hand, the permeability coefficient for helium was pressure independent, at least up to 4.4 atm, over the entire temperature range investigated. Hence such behavior may depend also on the molecular size of the penetrant. The anomalous permeation observed with some penetrant–membrane systems below the glass-transition temperature of the polymer has been discussed in some detail by Park.[12] It is

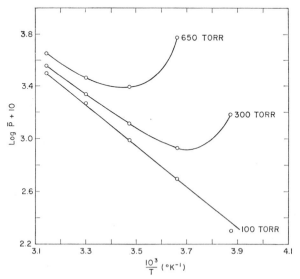

Figure 6. Permeability coefficients for methyl bromide in polyethylene—dependence of permeability coefficient on temperature and pressure.[40]

advisable to avoid any anomalous and unpredictable behavior in gas permeation processes by selecting operating temperatures well above such transition points.

For other types of systems, the shape of the log $\bar{P}$ versus $1/T$ plots depends on the relative variation of the diffusion and solubility coefficients with temperature. Interesting behavior is illustrated in Figure 6 for the methyl bromide–polyethylene system,[40] for which the diffusion coefficient is concentration dependent and the absorption isotherms deviate from Henry's law. The Arrhenius plot for $\bar{P}$ at a constant pressure at 100 torr is linear, as in the previous examples. The isobars at higher pressures pass through a marked minimum as the temperature is lowered, which is explained by the opposite temperature dependence of $\bar{D}$ and S for this system, as shown in Figure 7.[40]

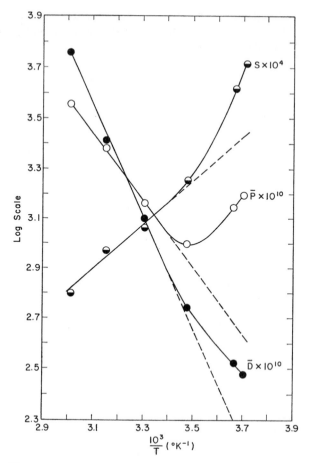

Figure 7. Permeability, diffusion, and solubility coefficients for methyl bromide in polyethylene at 650 torr.[40]

A qualitative picture of the simultaneous effects of pressure and temperature on $\bar{P}$ can now be formed.[27, 30] Penetrant–membrane systems in Category III often yield isothermal plots of log $\bar{P}$ versus applied penetrant pressure that are strongly nonlinear; the dependence of $\bar{P}$ on the pressure can be expressed for a number of systems by relations such as Equation 25. For penetrants that become less soluble in the polymer as the temperature is raised, that is, that exhibit exothermal heats of solution, the constants α and σ in Equation 25 tend to decrease with increasing temperature. Moreover, the limited information available indicates that σ decreases more rapidly than α

under these circumstances. When $\sigma \to 0$ and the solubility of the penetrant obeys Henry's law, Equation 25 reduces to Equation 27, which describes the pressure dependence of $\bar{P}$ for a number of systems in Category II. Isothermal plots of log $\bar{P}$ versus penetrant pressure are then linear, with positive slopes that decrease with increasing temperature. As the temperature is further raised and $\alpha \to 0$, $\bar{P}$ should become independent of pressure at any specific temperature. This behavior is characteristic of systems in Category I and has been termed as "ideal," because the solubility then obeys Henry's law and the diffusion coefficient is independent of pressure.

The above phenomenology, as well as recent theoretical considerations,[30] suggest that the permeability coefficients for a given system may exhibit a temperature and pressure dependence characteristic of *any* of the different categories of penetrant–membrane systems. The type of dependence observed is determined by the experimental temperature and pressure, as illustrated in Figure 8 for the carbon dioxide–polyethylene system.[27] The transition from one type of behavior to another, as described above, is clearly evident, and $\bar{P}$ is seen to become independent of pressure above the critical temperature of

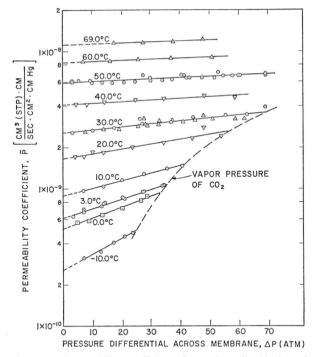

Figure 8. Permeability coefficients for carbon dioxide in polyethylene—dependence of permeability coefficient on temperature and pressure differential across membrane.[28]

the penetrant. Other penetrant–membrane systems show similar trends.[27, 30] However, no solubility and diffusivity data are available over sufficiently wide ranges of pressure and temperature for either the carbon dioxide–polyethylene system, or for other systems, and the validity of the generalizations above cannot be ascertained at this time.

In all the above cases the permeability coefficients increased with increasing penetrant pressure. A slight but significant *decrease* in $\bar{P}$ with increasing pressure has been observed for the helium–, nitrogen–, and fluoroform–polyethylene systems.[27, 30] This phenomenon is particularly interesting in the case of the first two systems, for which pressure-independent coefficients were expected for the reported experimental conditions. Stern, Fang, and Frisch[30] have shown the decrease in permeability coefficients to be a consequence of a decrease in the "free volume" of the penetrant–polymer systems with rising hydrostatic pressure. Since the "free volume" increases with increasing volume-fraction of dissolved penetrant, the above phenomenon reveals itself only when the penetrant solubility in the polymer is very low.

D.   Effect of Other Factors

Many attempts have been made in recent years to relate diffusion and solubility coefficients to various characteristic properties of penetrants and polymers. These properties have included the size, shape, polarity, and critical temperature of the penetrants, and the chemical structure, morphology, molecular weight, thermal expansivity, and glass-transition temperature of the polymers.[10–12] A number of interesting correlations have been discovered from such considerations, but their practical usefulness is limited.

Considerable efforts have also been made to explain the temperature and concentration dependence of the diffusion and solubility coefficients with the aid of physicochemical theory. As has been pointed out by Barrer,[16] diffusion coefficients are essentially rate constants and can be interpreted by the methods of chemical kinetics and irreversible thermodynamics. Diffusion may also be related to other types of rate processes, such as dielectric relaxation or viscous flow. Solubility is an equilibrium property characteristic of the penetrant–polymer system, and is amenable to thermodynamic and statistical thermodynamic analysis. Some of these concepts have proved valuable in providing an insight into the mechanism of gas and vapor transport in and through membranes, and the interested reader is referred to the extensive reviews on the subject.[10–12] The designer of separation processes cannot rely, as yet, on the predictions of these theories.

E. Permeation of Mixtures

Information on the permeability of polymeric membranes to mixtures of gases and vapors is limited. When the components of a mixture are gases that belong to Category I, according to the classification of Section II-A, it can be assumed that no interactions occur between them in the course of the permeation process. Each component permeates unperturbed by the presence of the other components, and its steady-state rate of permeation can be described by Equation 10. However, the equation must be expressed in terms of the *partial* pressure difference of the component across the membrane. The permeability coefficient for each component of the mixture will be the same as that for the component in its pure state and, in accordance with Equation 10, will be independent of partial or total pressures. This behavior is a consequence of the very low solubility of gases of Category I in many polymers.

The permeation of a number of gas mixtures at moderate pressures was found to conform to the assumptions made above: CO_2–H_2–O_2–N_2 through cellulose acetate–butyrate, polyethylene, and Trithene B (polychloro-trifluoroethylene);[7] NH_3–N_2–H_2 through polyethylene and Trithene B;[7] SO_2–O_2–N_2 through polyethylene;[7] O_2–N_2 and CO_2–O_2 through silicone rubber;[8, 9] O_2–N_2 through ethyl cellulose;[6] He–CH_4–N_2 (natural gas) through Teflon FEP (copolymer of tetrafluoroethylene and hexafluoropropylene);[26] and H_2–C_2H_6 through polypropylene.[41]

When the components of a gas mixture are more soluble in the membrane and Henry's law is not obeyed, it is no longer possible to assume that these permeate independently of each other. Nor can it be assumed, as in Equation 10, that the permeability coefficients are independent of pressure.[27] Rogers has summarized some of the results reported in the literature on the permeation of mixed gases and vapors.[11] In general, the more soluble component (or components) of a mixture tends to plasticize the membrane if its concentration therein is sufficiently high. The rates of permeation of some, or all, of the components are then increased to varying extents, and in extreme cases by orders of magnitude. This plasticizing effect depends mainly on the volume fraction of the plasticizing component and to much lesser degree on its chemical nature.[11] The *relative* increase in the rates of permeation of the different components may also vary or may remain constant; both types of behavior have been observed.

It should be evident from this summary that when a new separation process is being studied it is prudent to rely only on experimental permeability data obtained over the range of pressures, temperatures, and concentrations of interest.

III. ENGINEERING CONSIDERATIONS

A. General Principles

The separation of gases and vapors by the process of selective permeation is usually effected in one or more elementary separation units or devices, commonly called *stages*. These units may be visualized, in principle, as enclosures divided by means of a polymeric membrane in two sections. In a continuous process, a stream of the gas mixture to be separated is fed at an appropriately high pressure into one section of the stage. A fraction of the feed stream is allowed to permeate through and across the membrane into the opposite section of the stage, which is maintained at a lower pressure. As a result of this process, the feed is separated into two product streams: a low-pressure (permeated) stream enriched in the more permeable components of the feed, and a high-pressure (unpermeated) stream depleted in these components.

The degree of separation achievable in a single permeation stage, for any specific feed composition and membrane, depends on the operating variables. These include the pressures on the two sides of the membrane; the temperature; the fraction of the feed allowed to permeate, or "stage cut"; and the flow pattern of the gas streams on the two sides of the membrane. The flow pattern depends, in part, on the geometry of the stage. When the degree of separation obtained in a single stage does not meet requirements, several stages are connected in series to form a *cascade*. The conditions leading to the multiplication of the elementary separation effect in a permeation cascade are discussed in a later section. A gas permeation process resembles in many respects the well-known gaseous diffusion method for the separation of uranium isotopes, but the mechanism of gas transport through nonporous polymeric membranes is fundamentally different from that involved in the transport of gases through the porous barriers used in gaseous diffusion.

The "heart" of any permeation process is the membrane that will perform the desired separation. If the permeation process is to compete with conventional separation techniques from an economic standpoint, the membrane must exhibit the following properties:

1. A high permeability towards a specific component (or components) to be separated from a gas mixture.
2. A high selectivity for this component, that is, a high permeability relative to the other components of the mixture.
3. Chemical inertness and physical stability.
4. Continuity, that is, absence of pinholes or other mechanical defects.

A high permeability is necessary to minimize the membrane area requirements, hence the size of the permeation plant and the capital investment costs. A high selectivity is desirable to reduce the number of separation stages and the operating costs. The actual magnitude of the permeability and selectivity must be determined for each separation process from economic considerations. The desirability of chemical inertness and physical stability is self-evident. Membrane continuity is required in order to obtain the maximum selectivity characteristic of the membrane. The presence of pores or pinholes in the membrane will greatly reduce, and even destroy, the selectivity for the desired components, because of Knudsen flow or other types of gas transport through such imperfections. A theory of gas permeation by simultaneous diffusion and convection through membranes possessing pores has been developed by Frisch.[43]

The permeability of membranes to pure gases and gaseous mixtures has been discussed in the preceding section. The selectivity of a membrane for a component, A, of a mixture, relative to another component, B, is customarily expressed by an *ideal separation factor*, α^*_{A-B}, which is defined as the ratio of the permeability coefficients for pure A and B:

$$\alpha^*_{A-B} = \frac{\bar{P}^A}{\bar{P}^B} \tag{43}$$

If the membrane exhibits a higher selectivity for component A than for B, then $\alpha^*_{A-B} > 1$, whereas the opposite is true if $\alpha^*_{A-B} < 1$. No separation can occur, of course, for $\alpha^*_{A-B} = 1$. The ideal separation factor is, therefore, a separation index similar to the relative volatility used in fractional distillation. Michaels[44] has summarized the general selectivity characteristics of polymeric membranes as follows:

1. A membrane is selectively permeable toward that component of a gaseous mixture that has the highest critical temperature, the smallest molecular diameter, or both.
2. The selectivity of a membrane invariably decreases with increasing temperature.
3. Stiff-chain polymer membranes, although invariably less permeable to gases than flexible-chain polymers of similar chemical constitution, are more selective toward small molecules relative to large ones, or towards unsymmetrical relative to symmetrical molecules of equal size.

Some of these characteristics are illustrated in Table 1, which lists permeability coefficients for helium, nitrogen, and methane in a variety of membranes, and ideal separation factors for helium–nitrogen and helium–methane mixtures. This information is useful for the development of a

Table 1. Permeabilities and Separation Factors for Helium, Nitrogen, and Methane in Various Polymeric Membranes[26]

Membrane	Temp. (°C)	Permeability Coefficient, $\bar{P}\left(\dfrac{cm^3(STP)\cdot cm}{sec\cdot cm^2\cdot cm\ Hg}\right)$			Ideal Separation Factor	
		He	N_2	CH_4	$\alpha^*_{He-N_2}$	$\alpha^*_{He-CH_4}$
Silicone rubber	30.0	2.3×10^{-8}	1.5×10^{-8}	5.9×10^{-8}	1.5	0.39
Phenylene silicone rubber	30.0	1.5×10^{-8}	0.40×10^{-8}	2.0×10^{-8}	3.8	0.75
Nitrile silicone rubber	30.0	0.79×10^{-8}	0.21×10^{-8}	1.0×10^{-8}	3.8	0.79
Polycarbonate	30.0	6.7×10^{-9}	0.46×10^{-9}	0.36×10^{-9}	15	19
Teflon FEP	30.0	6.2×10^{-9}	0.25×10^{-9}	0.14×10^{-9}	25	44
Natural rubber	30.0	3.6×10^{-9}	1.05×10^{-9}	—	3.4	—
Polystyrene	30.0	3.5×10^{-9}	0.22×10^{-9}	0.23×10^{-9}	16	15
Trithene B	30.0	3.4×10^{-9}	0.012×10^{-9}	0.0084×10^{-9}	280	400
Ethyl cellulose	30.0	3.1×10^{-9}	0.28×10^{-9}	0.64×10^{-9}	11	4.9
Ethylene–vinyl acetate	30.0	2.1×10^{-9}	0.28×10^{-9}	1.1×10^{-9}	7.5	1.9
Viton A	30.0	1.7×10^{-9}	0.031×10^{-9}	0.016×10^{-9}	55	110
Polyvinyl chloride (plasticized)	30.0	1.4×10^{-9}	—	0.2×10^{-9}	—	7
Polyethylene	30.0	1.0×10^{-9}	0.19×10^{-9}	—	5.3	—
Polyvinyl fluoride	30.0	1.8×10^{-10}	0.019×10^{-10}	0.0065×10^{-10}	95	280
Mylar	25	1.0×10^{-10}	0.006×10^{-10}	0.006×10^{-10}	170	170
Saran	25	6.6×10^{-12}	0.018×10^{-12}	0.025×10^{-12}	370	260

permeation process for the recovery of helium from natural gas. Table 1 shows that most membranes are more permeable to helium than to the other gases, except for silicone rubber membranes, which are more permeable to methane. The helium–nitrogen and helium–methane selectivities reflect the relative magnitudes of the diffusivities and solubilities of these gases in the various membranes at the stated temperatures. If any trend can be discerned at all in Table 1, it is that the selectivities tend to increase with decreasing membrane permeability. Nature appears to be unkind to the designer of separation processes, who must select the appropriate separation membrane on the basis of a compromise between permeability and selectivity.

B. Membrane Separation Parameters

1. The Single Stage

Analytical studies on the separation of gas mixtures in a single permeation stage have been reported by a number of investigators, starting with Weller and Steiner.[5-7, 26, 45-51] These investigators have developed methods of calculating the extent of separation that can be obtained in such a stage and the corresponding membrane area requirements, as a function of the operating variables mentioned previously. Most of the work in this field has been reviewed recently by Kammermeyer[50] and by Stern and Walawender,[51] who also described various computer programs for the numerical evaluation of the proposed analytical methods.

In general, the analyses assume that the individual components of permeating gas mixtures do not interact with each other, and that the permeability coefficients for these components are independent of pressure. Moreover, with perhaps one exception, idealized or limiting gas flow conditions are considered inside the stage. Several examples of such analyses are presented below. The nomenclature used in this field is not fully consistent with that encountered in basic permeation studies.

The Perfect Mixing Assumption. In one of the limiting cases studied, Weller[5, 6] assumed that the gas on the high-pressure side of the stage (or membrane) is mixed so rapidly that its composition *at all points in the stage* is virtually the same as that of the unpermeated gas leaving the stage. The same assumption is made for the low-pressure side of the stage. These conditions of "perfect mixing" are shown diagrammatically in Figure 9, which is self-explanatory.

The rates of permeation of the components A and B of a binary gas

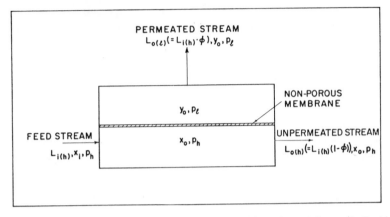

Figure 9. Diagram of a single permeation stage with perfect mixing on both sides of the membrane. $L_{i(h)}, L_{o(h)},$ and $L_{o(l)}$ = molar flow rates of feed, unpermeated, and permeated streams, respectively; p_h and p_l = total pressures on high- and low-pressure sides of stage, respectively ($p_h > p_l$); $x_i, x_o,$ and y_o = concentration (in mole-fractions) of any component in feed, unpermeated, and permeated streams, respectively; $\phi = L_{o(h)}/L_{i(h)}$ = stage "cut."

mixture are described for this case by the following forms of Equation 10:

$$y_o^A L_{o(l)} = P_0^A \left(\frac{A}{\delta}\right)(p_h x_o^A - p_l y_o^A) \tag{44}$$

$$(1 - y_o^A)L_{o(l)} = P_0^B \left(\frac{A}{\delta}\right)[p_h(1 - x_o^A) - p_l(1 - y_o^A)] \tag{45}$$

In the above equations, $L_{o(l)}$ is the molar flow rate of the permeated stream; p_h and p_l are the total pressures on the two sides of the membrane; y_o^A and x_o^A are the mole-fractions of the more permeable component (assumed to be A) in the permeated and unpermeated product streams, respectively; and the other symbols are as used previously. In view of the perfect mixing assumption, y_o^A and x_o^A are also the mole-fractions of A in the high- and low-pressure streams *inside* the stage, respectively.

The following material balances can also be written for the stage:

$$L_{i(h)} = L_{o(h)} + L_{o(l)} \tag{46}$$

and

$$L_{i(h)}x_i^A = L_{o(h)}x_o^A + L_{o(l)}y_o^A \tag{47}$$

where $L_{i(h)}$ and $L_{o(h)}$ are the molar flow rates of the feed and unpermeated streams, respectively; and x_i^A is the mole-fraction of component A in the feed. Let also

$$\alpha^* = \frac{P_0{}^A}{P_0{}^B} \tag{48}$$

$$\beta = \frac{L_{o(l)}}{\left(\dfrac{A}{\delta}\right)P_0{}^A} \tag{49}$$

and

$$\gamma = \frac{L_{o(h)}}{L_{i(h)}} \tag{50}$$

where α^* is the ideal separation factor. Equations 44–50 yield, by means of algebraic manipulations, the following expressions for $y_o{}^A$, $x_o{}^A$, and $x_i{}^A$, the mole-fractions of the more permeable component in the permeated, unpermeated, and feed streams, respectively:

$$y_o{}^A = \frac{\alpha^* - \left(\dfrac{p_h - p_l}{\beta}\right)}{\alpha^* - 1} \tag{51}$$

$$x_o{}^A = \left(\frac{\beta + p_l}{p_h}\right)\left[\frac{\alpha^* - \left(\dfrac{p_h - p_l}{\beta}\right)}{\alpha^* - 1}\right] = \frac{(\beta + p_l)y_o{}^A}{p_h} \tag{52}$$

$$x_i{}^A = \left[(1 - \gamma) + \frac{\gamma(\beta + p_l)}{p_h}\right]\left[\frac{\alpha^* - \left(\dfrac{p_h - p_l}{\beta}\right)}{\alpha^* - 1}\right]$$

$$= (1 - \gamma)y_o{}^A + \gamma x_o{}^A \tag{53}$$

Equation 53 can also be written in terms of the stage "cut," ϕ, since by definition

$$\phi = \frac{L_{o(l)}}{L_{i(h)}} \tag{54}$$

and

$$\gamma = 1 - \phi \tag{55}$$

Generally, the values of $P_0{}^A$ and $P_0{}^B$, γ or ϕ, p_h and p_l, $x_i{}^A$, and $L_{i(h)}$ are known. Then β is obtained from Equation 53, and the values of $y_o{}^A$ and $x_o{}^A$ are calculated from Equations 51 and 52. Finally, the membrane area A is calculated from Equation 49 for a specified thickness δ.

Weller's theory is limited to the separation of binary mixtures, but was extended by Huckins and Kammermeyer[45, 46] and by Brubaker and Kammermeyer[7] to ternary and quaternary mixtures. Another analytical solution of the perfect mixing case, which is particularly useful for multicomponent mixtures, is obtained by a simple iteration method.[26] This method is discussed below relative to a ternary mixture of components A, B, and C.

The problem is to determine the following unknown quantities given in Figure 9:

$$y_o{}^A, y_o{}^B, y_o{}^C; \quad x_o{}^A, x_o{}^B, x_o{}^C; \quad L_{o(l)} \quad \text{or} \quad L_{o(h)}; \quad \text{and } A$$

knowing the values of

$$x_i{}^A, x_i{}^B \quad \text{or} \quad x_i{}^C; \quad L_{i(h)}; \quad \phi; \quad p_h \quad \text{and} \quad p_l; \quad P_0{}^A, P_0{}^B, P_0{}^C; \quad \text{and } \delta.$$

The eight unknowns can be found by solving a set of eight simultaneous equations; these include three continuity equations similar to Equation 44 for the components A, B, and C, three material balance equations similar to Equation 47, and the two conditions:

$$\sum_n x_i{}^n = 1 \tag{56}$$

and

$$\sum_n y_o{}^n = 1 \tag{57}$$

where n designates the components.

The relations above yield the following expression for the membrane area:

$$A = \frac{L_{o(l)} y_o{}^A \delta}{P_0{}^A \left[\dfrac{p_h}{1 - \phi} (x_i{}^A - \phi y_o{}^A) - p_l y_o{}^A \right]} \tag{58}$$

The mole-fractions $y_o{}^B$ and $y_o{}^C$ in the permeated stream are given by

$$y_o{}^B = \frac{\left(\dfrac{p_h x_i{}^B}{1 - \phi} \right)}{\left(\dfrac{L_{o(l)} \delta}{P_0{}^B A} \right) + \left(\dfrac{\phi}{1 - \phi} \right) p_h + p_l} \tag{59}$$

and

$$y_o{}^C = \frac{\left(\dfrac{p_h x_i{}^C}{1 - \phi} \right)}{\left(\dfrac{L_{o(l)} \delta}{P_0{}^C A} \right) + \left(\dfrac{\phi}{1 - \phi} \right) p_h + p_l} \tag{60}$$

The following procedure is used to solve these equations:

1. A value is assumed for $y_o{}^A$.
2. The membrane area is calculated from Equation 58, and $L_{o(h)}$ or $L_{o(l)}$ is calculated from the material balance in Equation 46.

3. Next, $y_o{}^B$ and $y_o{}^C$ are calculated from Equations 59 and 60, respectively.
4. $\sum_n y_o{}^n$ is determined, and steps 1–3 are repeated until $\sum_n y_o{}^n = 1$.
5. $x_o{}^A$, $x_o{}^B$, and $x_o{}^C$ are calculated from the material balance equations.
Relations 58, 59, and 60 can be written also in terms of $L_{i(h)}$, $L_{o(h)}$, and $L_{o(l)}$.

An analysis of membrane separation parameters based on the perfect mixing assumption has been described recently, with special reference to air separation.[51]

The "No Mixing" Assumption. Another type of flow pattern that has been studied by several investigators can be designated as "cross flow": the gas stream on the high-pressure side of the stage flows parallel to the membrane,

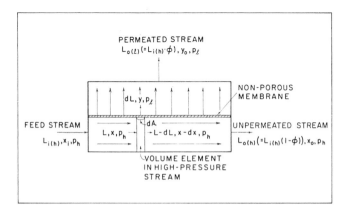

Figure 10. Diagram of a single permeation stage for cross flow and with no mixing on either side of the membrane. dL = molar permeation rate through element of area dA; L = local molar flow rate on high-pressure side of stage; x and y = local concentration (in mole-fractions) of any component on high- and low-pressure sides of stage, respectively. Other symbols are as in Figure 9.

whereas the permeated stream flows perpendicular to, and away from, the membrane. It is assumed that no mixing takes place on the high-pressure side, as would be expected for plug (piston) flow. It is also assumed that no mixing occurs on the low-pressure side. The composition of the permeated gas at any point near the membrane is then determined by the relative rates of permeation of the feed components at that point.

The physical conditions in the permeation stage are illustrated for this case in Figure 10. The composition of the permeated gas leaving the stage, the stage cut, and the required membrane area are computed as follows. The *local* permeation rates of the components of a binary mixture, at any cross section of the stage, can be expressed by the following relations for steady-state conditions:

$$-y^A \, dL = P_0^A \left(\frac{dA}{\delta}\right)(p_h x^A - p_l y^A) \tag{61}$$

$$-(1 - y^A) \, dL = P_0^B \left(\frac{dA}{\delta}\right)[p_h(1 - x^A) - p_l(1 - y^A)] \tag{62}$$

where x^A and y^A are the local composition (in mole-fractions) of component A on the high- and low-pressure sides of the membrane, respectively; dL is the total (molar) permeation rate through an element of membrane area dA; and the other symbols are as used before.

The average composition of the permeated gas stream leaving the stage can be obtained in terms of the mole-fraction of component A from

$$y_o^A = \frac{L_{o(l)}^A}{L_{o(l)}} = \frac{L_{i(h)}^A - L_{o(h)}^A}{L_{i(h)} - L_{o(h)}} = \frac{L_{i(h)}^A - L_{o(h)}^A}{(L_{i(h)}^A + L_{i(h)}^B) - (L_{o(h)}^A + L_{o(h)}^B)} \tag{63}$$

$L_{i(h)}^A$ and $L_{i(h)}^B$ are the molar flow rates of components A and B in the feed stream, and $L_{o(h)}^A$ and $L_{o(h)}^B$ are the corresponding flow rates in the unpermeated stream leaving the stage. The values of $L_{i(h)}^A$ and $L_{i(h)}^B$ are known from the composition and flow rate of the feed stream, and $L_{o(h)}^A$ and $L_{o(h)}^B$ can be calculated by the method of Weller.[5, 6] Starting with Equations 61 and 62, Weller derived by an exact analytical procedure the relation

$$\ln \frac{L^B}{L_{i(h)}^B} = R \ln \left[\frac{u_i - (E/D)}{u - (E/D)}\right] + S \ln \left[\frac{u_i - \alpha^* + F}{u - \alpha^* + F}\right]$$
$$+ T \ln \left[\frac{u_i - F}{u - F}\right] \tag{64}$$

where L^B is the *local* flow rate of component B on the high-pressure side of the stage, and the quantities D, E, F and R, S, T depend only on the pressure ratio p_h/p_l $(= r)$ and the ideal separation factor α^*, that is, on P_0^A and P_0^B; definitions of these quantities are given in Appendix A. The function u above is defined by

$$u \equiv -Di + (D^2 i^2 + 2Ei + F^2)^{1/2} \tag{65}$$

where $i = L^A/L^B$, and L^A is the flow rate of A corresponding to L^B; u_i is the value of u at the stage inlet, where $i = i_i = L_{i(h)}^A/L_{i(h)}^B$. Thus, L^B can be calculated for any point in the stage as a function of i, for a specified r, α^*, and feed composition and flow rate. To find $L_{o(h)}^B$ from Equation 64, a value must be *assumed* for i at the stage outlet, that is, for i_o. The corresponding $L_{o(h)}^A$ is then obtained by virtue of the relation $L_{o(h)}^A = i_o \, L_{o(h)}^B$. The composition of the permeated gas can now be calculated from Equation 63, and the stage cut ϕ is obtained from

$$\phi = \frac{L_{o(l)}}{L_{i(h)}} = \frac{L_{i(h)} - L_{o(h)}}{L_{i(h)}} = \frac{(L_{i(h)}^{A} + L_{i(h)}^{B}) - (L_{o(h)}^{A} + L_{o(h)}^{B})}{(L_{i(h)}^{A} + L_{i(h)}^{B})} \quad (66)$$

The value of y_o^A for a *specified* stage cut must be obtained by trial-and-error. Finally, the required membrane area, A, is calculated from

$$A = -\frac{\delta}{P_0^B} \int_{i_i}^{i} \frac{L^B \, di}{(f - i)[(p_h/1 + i) - (p_l/1 + f)]} \quad (67)$$

where $f = y^A/(1 - y^A)$, which is the ratio of Equations 61 and 62. The integral can be evaluated numerically or graphically.

Another method of determining the separation and membrane area under no-mixing conditions has been described by Naylor and Backer.[47] These investigators derived the following expression for y_o^A, which is sometimes termed the "enrichment":

$$y_o^A = (x_o^A)^{-1/\epsilon} \left(\frac{1 - \phi}{\phi}\right)$$

$$\times \left\{(1 - x_o^A)^\sigma \left[\frac{(1 - \phi)x_o^A + \phi y_o^A}{1 - [(1 - \phi)x_o^A + \phi y_o^A]}\right]^\sigma - (x_o^A)^\sigma\right\} \quad (68)$$

here, $\sigma = (\epsilon + 1)/\epsilon$; $\epsilon (= \alpha - 1)$; and α is the *local or point separation factor*. As shown in Appendix A, α depends on α^*, the pressure ratio r, and the local composition x^A. Naylor and Backer assumed that the dependence on x^A was negligible for any one stage, and used conservatively a value of α corresponding to x_o^A. Hence y_o^A can be determined from Equation 68 for a specified α^*, r, and ϕ, and assumed values of x_o^A. The internal consistency of the calculations can be checked by means of the material balance of Equation 46, which can be written in the form:

$$y_o^A = \frac{1}{\phi} [x_i^A - (1 - \phi)x_o^A] \quad (69)$$

Naylor and Backer have not calculated membrane areas, but these can be obtained from Equation 61. Rearrangement and integration of this equation yields:

$$A = \int_{x_o^A}^{x_i^A} \frac{\delta y^A \, dL}{P_0^A (p_h x^A - p_l y^A)} \quad (70)$$

The evaluation of this integral is discussed in Appendix B. The usefulness of the Naylor and Backer method is limited to smaller values of α because of the assumption that α does not vary with changing composition across the stage.[51] Hwang[52] has pointed out that this method becomes identical with the Weller method for no-mixing conditions when $\alpha = \alpha^*$. As shown in Appendix A, this occurs when $r = \infty$ ($p_l = 0$).

The iteration method used previously for perfect mixing can be applied also to the case of no mixing.[26, 51] The stage is divided in a large number of hypothetical sections, with the assumption of perfect mixing still holding for each individual section. A small incremental value is then taken for the molar flow rate $L_{o(l)}$, and the values of $L_{o(h)}$; A; y_o^A, y_o^B, and so on; and x_o^A, x_o^B, and so on, are computed as described earlier. The unpermeated gas stream leaving the first section is assumed to become the feed to the second section, and the calculation is repeated until the desired degree of removal of the more permeable component from the feed is achieved.

Computer programs for all three methods have been described.[50, 51] The values of the enrichment and the membrane area calculated by the Weller and iteration methods are in good agreement, at least for the cases studied.[51] The Naylor and Backer method should be used only with due regard to its limitations. The iteration method may require shorter computing times when the stage cut is specified and, in particular, for multicomponent separations.

Other Stage Flow Patterns. Several other stage flow patterns and their effect on the achievable enrichment have been studied. Oishi et al.[49] investigated the important cases when the high- and low-pressure gas streams flow either cocurrently or countercurrently, with no mixing taking place on either side of the membrane. The analysis of these cases is more complex and the basic transport equations cannot be solved analytically. Breuer and Kammermeyer[48] examined the effect of concentration gradients parallel and perpendicular to the membrane on the idealized cross flow discussed earlier.

The question as to which of these flow patterns is more realistic can be answered only in relation to the actual geometry of the permeation stage used in practice. Countercurrent flow with no mixing is one of the most efficient flow patterns in terms of achievable separation, whereas perfect mixing is the least efficient.[49] Other conditions yield separations that lie between the values obtained in these extreme cases. When the ideal separation factor and pressure ratio are small, the differences between some of the cases studied are relatively slight. This is illustrated in Figure 11, which compares enrichments and membrane areas calculated for perfect mixing and cross flow with no mixing, for two different values of α^* and r. The illustration refers to the separation of about 123 tons of air/day (air is assumed to be a binary mixture of 20.9 mole-% O_2 and 79.1 mole-% N_2) by means of a hypothetical membrane with a permeability to oxygen similar to that of silicone rubber. This writer feels that the perfect mixing assumption is useful

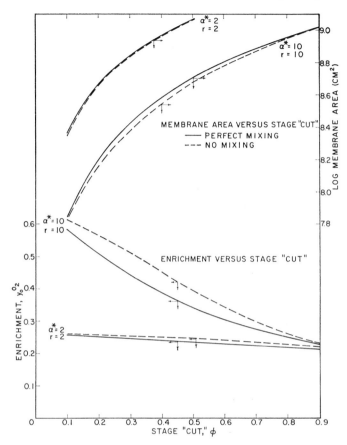

Figure 11. Comparison between perfect mixing and "no-mixing" conditions. Dependence of membrane area and enrichment on the stage "cut," as a function of the ideal separation factor, α^*, and the pressure ratio, r.[51] Feed composition: air (20.9 mole-% O_2 and 79.1 mole-% N_2). Feed rate: 1×10^6 cm³ (STP)/sec ($\sim$123 tons/day). Type of membrane: hypothetical. Membrane thickness: 2.54×10^{-3} cm (1 mil). Permeability coefficients: $P_{O_2} = 5 \times 10^{-8}$ cm³ (STP) · cm/sec · cm² · cm Hg, P_{N_2} variable. Pressures: p_h variable, $p_l = 38$ cm Hg.

for preliminary evaluations of new separation processes because of its inherent simplicity, and also because it yields conservative values for the achievable enrichment and required area of membrane.

All the previous discussions referred to steady-state permeation. Barrer[16] has considered the possibility of gas separation in the *transient state*. His calculations show that very high separations can be obtained in the early stages of permeation, before much of the less permeable component has passed through the membrane. The practical application of such a process is difficult to visualize.

2.   The Permeation Cascade

The design principles of a cascade of permeation stages are identical to those of a gaseous diffusion cascade for the separation of uranium isotopes. These principles were developed during World War II and are outlined in the classical monograph of K. Cohen[53] The mathematical model of a cascade with a nonporous membrane for its separation element differs from one utilizing a porous (gaseous diffusion) barrier only in the characterization of

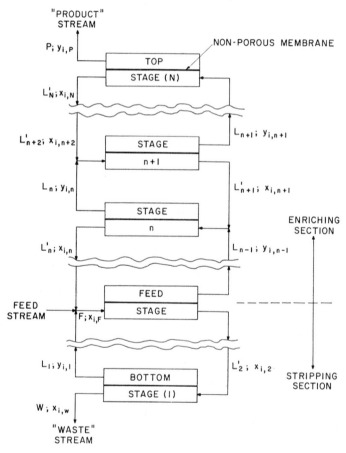

Figure 12a. Diagram of a permeation cascade.

its separation factor. Principles of cascade design, as applied to gas permeation, have been discussed by Naylor and Backer,[47] Hwang and Kammermeyer,[54] and, more recently, by Blumkin.[55] The latter has used a simple procedure for calculating multicomponent concentration gradients developed in a cascade; this procedure is summarized below.

Blumkin has considered a cascade of the type shown in Figure 12a. The cascade has no recycle flow at the end stages, so that the "product" stream is equal to the permeated stream leaving the end stage at the top of the cascade, and the "waste" stream is equal to the unpermeated stream leaving the end stage at the bottom of the cascade. The terms "product" and "waste" are arbitrary, but according to convention the desired component is enriched in the product stream. The following nomenclature is adopted to index the various stages and gas streams shown in Figure 12a.

n = Stage index
i = Component index
N = Total number of stages
n_F = Stage receiving cascade feed
L = Upflowing (permeated) stream
L' = Downflowing (unpermeated) stream
F = Cascade feed rate
P = Product withdrawal rate
W = Waste withdrawal rate
y_i = Mole-fraction in upflowing stream
x_i = Mole-fraction in downflowing stream
$x_{i,F}$ = Mole-fraction in cascade feed stream
$y_{i,P}$ = Mole-fraction in cascade product stream
$x_{i,W}$ = Mole-fraction in cascade waste stream

Two types of relations are required to determine the cascade gradient and size:

1. Relations between the concentration of specified components in the upflowing (permeated) stream leaving a stage, $y_{i,n}$, and that in the downflowing (unpermeated) stream from the next higher stage, $x_{i,n+1}$. These are obtained from appropriate material balances and are frequently called "operating-line" equations.

2. Relations between the concentrations of specified components in the upflowing and downflowing streams from any single stage, $y_{i,n}$ and $x_{i,n}$, respectively. These are known as "equilibrium line" equations, although no equilibrium in the thermodynamic sense is involved here.[54]

The material balances for the cascade under consideration are as follows:
For $n > n_F$ (enriching section),

$$L'_{n+1} = L_n - P = L_n + W - F \tag{71}$$

$$x_{i,n+1} = \frac{L_n y_{i,n} + W x_{i,W} - F x_{i,F}}{L'_{n+1}} \tag{72}$$

For $n \leq n_F$ (stripping section),

$$L'_{n+1} = L_n + W \tag{73}$$

$$x_{i,n+1} = \frac{L_n y_{i,n} + W x_{i,W}}{L'_{n+1}} \tag{74}$$

Also, because there is no recycle flow from the end stages:

$$L_N = P \quad \text{and} \quad L'_1 = W$$

$$y_{i,N} = y_{i,P} \quad \text{and} \quad x_{i,1} = x_{i,W}$$

The "equilibrium" relation between $y_{i,n}$ and $x_{i,n}$ depends on the stage flow pattern. Such a relation can be obtained for perfect mixing conditions from the ratio of Equations 61 and 62, whereas for cross flow with no mixing it can take the form of Equation 68. For more complex flow conditions, the values of $y_{i,n}$ may have to be computed as a function of $x_{i,n}$ (or vice versa) by numerical methods. In isotope separation, the "equilibrium" relations are usually defined in terms of a *stage separation factor* defined by

$$\alpha_{i,n} = \frac{y_{i,n}/y_{k,n}}{x_{i,n}/x_{k,n}} \tag{75}$$

where $y_{k,n}$ and $x_{k,n}$ are the concentrations (in mole-fractions) of a key, or reference, component in the upflowing and downflowing streams from the stage, respectively. The stage separation factor for gas permeation can be a function of the stage cut, the ideal separation factor, and the pressure ratio. As shown by Blumkin,[55] it may be very difficult to derive an explicit relation for $\alpha_{i,n}$, in which case it is again necessary to resort to numerical techniques.

In a common situation, the magnitudes of two of the streams and one set of concentrations, such as F, P, and the $x_{i,F}$'s, are known. If the stage upflows (L_n's) and the waste concentrations ($x_{i,W}$'s) are specified or assumed, the cascade gradient can be calculated from the mass-balance and "equilibrium" equations. Thus (beginning at the bottom) the calculation is performed from stage to stage until a set of product concentrations is obtained. Specification of L_n's implies that the stage cuts are also specified. The results are tested by means of the material balance equations for the external streams of the cascade:

$$P + W = F \tag{76}$$

and
$$Py_{i,P} + Wx_{i,W} = Fx_{i,F} \tag{77}$$

The procedure is to calculate the $y_{i,P}$'s from Equations 76 and 77 using the assumed $x_{i,W}$'s. The values of $y_{i,P}$ from the cascade gradient calculation are then compared with those obtained from the material balance over the cascade. If these values differ by more than some small value selected to represent the convergence criterion, the calculations are repeated as often as necessary with new sets of $x_{i,W}$'s. Blumkin has pointed out that "the success in arriving at a balanced cascade for a multicomponent system depends upon the effectiveness of the method used in making re-estimates of starting values that are significantly closer to the solution set than any previous estimate." Blumkin has developed computer programs for cascade gradient calculations handling up to four components in a cascade up to 60 stages. Other cascade properties can be determined by the usual procedures. The above calculations become more difficult if stage membrane areas are specified rather than stage upflows.

If the ideal separation factors are small, hence the number of stages required to perform a desired separation is large, it may be useful to consider the design of an *ideal* cascade. Such a cascade is defined as one in which no mixing of streams of unequal concentrations takes place. In the nomenclature of Figure 12a, the condition

$$x_{i,n+1} = y_{i,n-1} \tag{78}$$

must be satisfied for each stage. The ideal cascade minimizes the sum of the upflows of all the stages and thus the volume of the equipment and the power requirements. However, it does not necessarily minimize the cost of the cascade. The cascade gradient is calculated in this case by the described procedure, with the additional restriction imposed by Equation 78. Iteration procedures must be used to satisfy this restriction. An example of a small ideal cascade for carbon dioxide control in sealed environments is given in the following section; the intended use of such a cascade in aerospace applications calls for the minimization of power requirements.

C. Permeator Design

The special apparatus or device containing the membrane, and in which the separation process actually takes place, is perhaps the most important mechanical component of a large permeation plant. The importance of this apparatus, which may be called a "permeator," arises from the very large membrane area necessary at present for any large-scale permeation process. Since the total size of permeator equipment can be assumed to be roughly proportional to the membrane area employed, it is apparent that the permeator cost may control the capital investment costs of the permeation process. Consequently, the economic success of the process may depend on the design of efficient permeators, which must fulfill the following requirements:

1. Support very thin membranes under the large pressure differentials required in large-scale permeation processes.

2. Enclose very large areas of membrane in small volumes to reduce the size of the permeators.

3. Cause a minimum pressure drop in the high- and low-pressure gas streams, because such pressure drops increase the operating costs.

4. Be constructed from inexpensive materials.

5. Be simple to fabricate and assemble.

The design of permeators will also be affected by the chemical properties of the gases to be separated and the selected operating conditions. Since the above requirements are interdependent, appropriate compromises must be made in design characteristics and choice of construction materials. In the final analysis, permeator design and performance are dictated by economic considerations, some of which are discussed in a later section.

Several types of permeators have been described in the technical and patent literature, and three of the more basic ones are reviewed below.

1. Spiral Permeators

The spiral design provides a simple and logical method of packing large membrane areas inside cylindrical pressure vessels. Osburn and Kammermeyer[56] built and successfully operated a small permeator of this type.

Their membrane was a flat polyethylene tubing wrapped around a perforated pipe that served as a gas collector. This assembly was contained inside a pressure chamber consisting of a larger steel pipe. Sheets of filter paper separated adjacent turns of tubing and also lined its interior to provide flow paths for the high- and low-pressure gas streams. The compressed feed mixture was introduced in the pressure chamber and circulated around the external surface of the wrapped tubing. A fraction of the mixture permeated through the wall of the tubing, then flowed along a spiral path in its interior until it reached the collector. Osburn and Kammermeyer separated air and a helium–oxygen mixture in this permeator.

The use of compressed filter paper to conduct gas streams along sizable lengths of plastic tubing, as would be required for large-scale applications, has the disadvantage of causing large pressure drops in these streams. A spiral permeator designed to decrease such uneconomical pressure drops has been described more recently by Michaels.[57] According to this design, a separation membrane in the shape of an elongated strip is wrapped together with a flexible base strip around a hollow cylindrical core (Figure 12b).

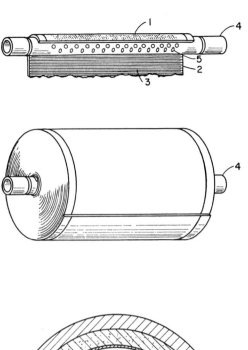

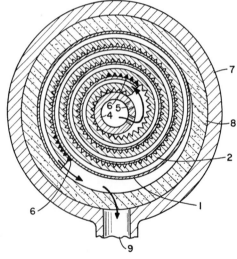

Figure 12b. Isometric views and radial section of spiral permeator.[57] Top: isometric view showing a partial turn of the membrane and base strip assembly wound on a hollow perforated core, at the start of the winding operation. Middle: isometric view of completed permeator. Bottom: radial section of completed permeator inside pressure vessel. 1, Membrane; 2, base strip; 3, shallow grooves parallel to axis of inner core; 4, hollow inner core; 5, perforations in inner core; 6, adhesive; 7, cylindrical housing; 8, porous material; 9, unpermeated gas outlet.

The base strip is provided on both sides with shallow grooves that serve as channels for gas flow. The grooves on the side facing the cylindrical core are parallel to the core axis, whereas the grooves on the opposite side are in a plane perpendicular to the axis. The whole assembly is contained in a pressure vessel, as illustrated on the bottom of Figure 12b. The reader is referred to Michaels' patent[57] for a description of the proposed sealing method and other constructional details.

In operation, the high-pressure gas mixture to be separated may be introduced into the inner core, whence it will flow through perforations in the core into the nearest end of the spiral membrane-and-base-strip laminate. As shown in Figure 12b, the mixture enters the grooves facing away from the inner core, follows a spiral and radial flow path to the other end of the laminate, and then leaves the permeator through an outlet in the outer pressure vessel. Along this path, a fraction of the mixture permeates the membrane, flows along the grooves parallel to the core axis, and then emerges from the permeator through outlets at its two extremities. An appropriate low pressure must be maintained in the latter grooves.

The membrane packing density achieved with this design depends on the thickness of the membrane-base strip laminate. For example, if the base strip is 0.050 in. thick, the grooves are 0.010 in. deep, and the membrane is 0.001 in. thick, about 240 ft² of membrane will be contained per cubic foot of total permeator volume. No pressure-drop measurements or mass-transfer characteristics have been reported for this permeator.

2.   *"Flat-Plate" Permeators*

Another concept of permeator design is modeled after the plate-and-frame construction of filter presses. One of the first permeators of this type was described in the patent literature by Steiner and Weller.[58] Their apparatus consists, in principle, of a series of side-by-side chambers, each of which is formed by sealing flat sheets of membrane to the two sides of a rectangular frame. A large number of such frames are clamped together inside a cylindrical pressure vessel. Adjacent membranes are separated by sheets of fibrous or porous materials, such as blotting paper, which serve to support the membranes under pressure and also to conduct the permeated gas. A stream of the gas mixture to be fractionated is introduced into the pressure vessel, and enters the assembly of chambers through special passages in one end of the frames. The gas in each of the chambers then flows along the sheets of membrane forming the chamber walls, and a fraction thereof permeates through the membranes. The permeated gas diffuses through the porous sheets in a direction parallel to the membrane surfaces and is withdrawn through a system of manifolded collector blocks. The fraction of the gas

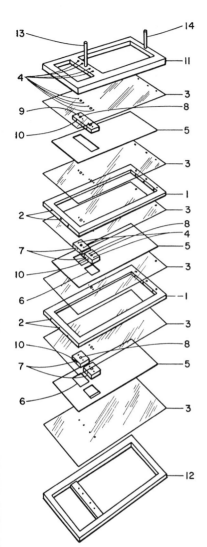

Figure 13. "Flat plate" permeator, exploded view.[58] 1, Rectangular frame; 2, gas passages in frame; 3, membrane; 4, holes in membrane, frame, or collector blocks for bolting; 5, porous sheet; 6, cut-out portion in porous sheet; 7, collector blocks; 8, passages in collector blocks; 9, uppermost collector block; 10, lateral passages in collector blocks; 11, upper clamping frame; 12, lower clamping frame; 13, outlet for permeated gas; 14, outlet for unpermeated gas.

mixture that does not permeate leaves the chambers through a manifold of passages and channels in the end of the frames opposite to the inlet passages. An exploded perspective view of the described assembly is shown in Figure 13. An apparatus of this type was constructed at the U.S. Bureau of Mines and found to operate satisfactorily for laboratory purposes.[59]

Large Weller-Steiner permeators will be subject to sizable pressure drops in the permeated gas streams, which are forced to diffuse inside long porous sheets under compression. This disadvantage can be minimized by supporting and separating adjacent membranes with more open structures, such as pairs of screens or honeycomb layers separated by spacers. An example of a plate-and-frame permeator embodying such structures has been described by Megibow[60] for use as an artificial lung or an artificial kidney. His apparatus could be adapted for the separation of gas mixtures, but its cost would be prohibitively high for this purpose.

A practical permeator that appears to fulfill the requirements outlined at the beginning of this section was developed by this writer and his associates.[26, 61] Their design evolved from the observation that thin polymeric membranes can be supported inexpensively on coarse metal screens, even

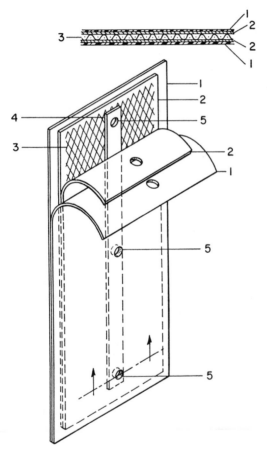

Figure 14. Permeation "septum": cross section and perspective.[61] 1, Membrane; 2, fibrous sheet; 3, screen; 4, metal strip; 5, holes drilled through assembly of septa.

under very high pressure differentials, if sheets of an appropriate fibrous material are inserted between membranes and screens. This observation led to the construction of a basic permeator element, called a "septum," which is shown in Figure 14. It consists of a rectangular piece of common metal flyscreen provided on both sides with thin strips of metal running lengthwise along its central axis. The screen, together with the metal strips, is enclosed in an envelope of fibrous material, which is inserted in turn in an outer envelope made of the separation membrane. The latter is formed by sealing the membrane around the edges of the screen. The fibrous sheet which supports the membrane must satisfy two conditions:

1. Its gas permeability in a direction *normal* to its surface must be at least an order of magnitude larger than the permeability of the separation membrane; the permeability, or porosity, parallel to the surface is unimportant, because the sheet does not have to conduct gases in that direction.

2. It must have sufficient strength to span the openings in the screen without appreciable sagging when a high pressure is applied to the membrane it cushions.

Fibrous materials that combine satisfactory porosity with structural rigidity include compressed paper of the kraft type and compressed fiberboard as used for electrical insulation. Blotter papers, filter papers, glass fiber papers, and monofilament woven organic materials can also be used, but require special precautions due to their softness.[61]

To construct a permeator, a large number of septa are stacked in parallel on the center strips, as shown in Figure 15. The strips are then cross-drilled at regular intervals and serve a dual purpose. First, the thickness of the strips separates adjacent permeation septa and provides space for the circulation of the high-pressure (unpermeated) gas stream. Second, by aligning the openings drilled through the metal strips in each septum, a manifold is formed for the removal of the low-pressure (permeated) gas stream (Figure 15, top). Sealing between septa at the manifold can be accomplished by gasketing, by heat sealing, or with epoxy cement; the membrane could serve as its own gasket if it is sufficiently thick and elastic. The assembly of stacked septa is held and compressed together by a number of high-tensile-strength bolts, which are inserted in some of the vertically aligned openings in the metal strips (Figure 15, bottom).

In operation, the stack of septa is contained inside a high-pressure vessel, so that the gas mixture to be separated flows parallel to the stack, over the surfaces of the septa. The fraction of gas permeating through the membranes flows across the supporting fibrous sheets, then along the screens inside the septa, and into the manifold formed by the cross-drilled metal strips. Tubular headers on top and bottom of the stack are provided to collect and remove the

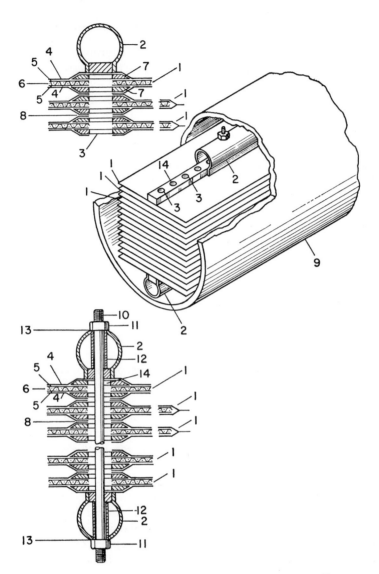

Figure 15. Assembly of permeation septa inside pressure vessel.[61] Top: upper section of a centrally-supported stack of permeation septa. Middle: stacked assembly of septa. Bottom: staying arrangement. 1, Septum; 2, manifold conduit; 3, vertically aligned cross-flow openings; 4, membrane; 5, fibrous sheet; 6, screen; 7, metal strip; 8, seal; 9, pressure vessel; 10, high-tensile-strength bolt; 11, nut; 12, sleeve; 13, O-ring seal; 14, vertically aligned opening for staying arrangement.

permeated gas. The gas inside the septa flows in a direction *perpendicular* to the metal strips. Therefore, its path to the manifold is short and the pressure drop inside the septa is low. This pressure drop can be further reduced, at the expense of membrane packing density, by using more than one screen inside the septa. When using two layers of flyscreen, 0.060 in. spacing of septa, and 0.001 in.-thick membranes, a packing density of about 240 ft² of membrane/ft³ is achieved.

Considerable work has been done to optimize the construction of the septa, which were tested in a permeation pilot plant over extended periods of time.[26, 61] A complete permeator with stacked septa has also been constructed with a number of minor modifications for use in the desalination of seawater by reverse osmosis. This permeator contained 50 ft² of anisotropic (Loeb-type) cellulose acetate membrane and was subjected to extensive tests.[62]

3. Hollow-Fiber Permeators

Several industrial organizations have introduced recently permeators consisting of bundles of small-diameter hollow fibers. Some of these permeators were developed specifically for the desalination of seawater,[63-65] but the same design concept can be used also for the separation of gases. The hollow fibers serve as separation barriers, in the place of plastic tubing or flat membranes. The configuration of a hollow-fiber permeator is similar to that of a "U" tube single-end heat exchanger. One end of the bundle of fibers is potted in a block of epoxy resin, which forms the equivalent of a tube sheet. The flow of the high-pressure unpermeated gas inside the permeator shell, and across the bundle of fibers, is countercurrent to the flow of the permeated gas inside the individual fibers.

The main advantage of such permeators is the tremendously large permeation area that can be packed per unit of equipment volume. It has been reported that areas of the order of 12,000 ft²/ft³ can be contained, using hollow fibers with an OD of 0.0018 in. (45 μ) and an ID of 0.00096 in. (24 μ). These fibers are thinner than a human hair. Another important advantage lies in the relatively high strength of some fibers, which makes it possible to reduce the wall thickness to less than 0.0005 in. and thus increase the rate of permeation. This inherent strength also eliminates the necessity for porous supports and greatly simplifies the construction of the permeator.

One of the problems that may be encountered in separating gases with this type of permeator is a large pressure drop inside the hollow fibers. Although no information is available on this subject, it may be possible to minimize the pressure drop by using fibers of low gas permeability. In view of the large permeation area available, this procedure should not greatly impair the efficiency of the permeator. Additionally, less permeable fibers may exhibit higher selectivity, as discussed in Section III-A.

The installed cost of a small hollow-fiber permeator has been stated to range between about \$0.40 and \$0.80/ft² of permeation area.[64] By comparison, the cost of septum-type permeators can be reduced to this range only for very large units.

IV.   EXAMPLES OF GAS PERMEATION PROCESSES

A variety of gas permeation processes of interest to industry or for environmental control have been proposed in the literature, and some of these have been discussed by Stern[51, 66] and, more recently, by Kammermeyer.[50] Four of the processes that have been analyzed in greater detail are reviewed in this section.

A.   Separation of Oxygen from Air

The separation of oxygen from air is one of the earliest applications proposed for the technique of selective permeation through nonporous membranes[3, 4], but it was not until 1950 that an economic evaluation of this process was undertaken by Weller and Steiner.[5, 6, 67] Their results illustrate well the hurdles still facing the development of a practical gas permeation process, although more effective membranes are presently available for air separation than considered by these investigators.

Weller and Steiner examined two different extents of separation: (a) a single-stage permeation process yielding only oxygen-enriched air (32.6 % O_2), and (b) a five-stage process producing oxygen of moderate purity (91.1 % O_2). The membrane selected by Weller and Steiner for this purpose was ethyl cellulose. The results of their economic evaluation for small plants, producing 120,000 ft³ of enriched air or oxygen/hr, are shown in Table 2. Only power requirements were calculated, since no large-scale permeation equipment had been devised at that time and consequently the cost of plant investment could not be determined. Weller and Steiner compared the power requirements of the two permeation processes with the corresponding requirements of the conventional cryogenic method of air separation. On this basis, they concluded that the production of oxygen by multistage permeation was not competitive. They felt, however, that the single-stage process for producing oxygen-enriched air was much more promising, because its power requirements were only 30 % higher than those of the cryogenic process.

Weller and Steiner's conclusions appear too optimistic in the light of present-day technology. First, the efficiency of low-temperature air separation

Table 2. Separation of Oxygen from Air[5,6]

Operating Conditions	Single-Stage Process	Five-Stage Process
Product composition	32.6 mole-% O_2	91.1 mole-% O_2
Plant size	120,000 ft^3 (product)/hr	120,000 ft^3 (product)/hr
Pressure of feed stream	8 atm (abs.)	8 atm (abs.)
Pressure of permeated stream	1 atm (abs.)	1 atm (abs.)
Temperature	30°C	30°C
Membrane composition	Ethyl cellulose	Ethyl cellulose
Membrane thickness	0.001 in. (1 mil)	0.001 in. (1 mil)
Permeability coefficients		
For oxygen $(P_0{}^{O_2})$	$9.6 \times 10^{-10} \dfrac{cm^3(STP) \cdot cm}{sec \cdot cm^2 \cdot cm\ Hg}$	$9.6 \times 10^{-10} \dfrac{cm^3(STP) \cdot cm}{sec \cdot cm^2 \cdot cm\ Hg}$
For nitrogen $(P_0{}^{N_2})$	$2.8 \times 10^{-10} \dfrac{cm^3(STP) \cdot cm}{sec \cdot cm^2 \cdot cm\ Hg}$	$2.8 \times 10^{-10} \dfrac{cm^3(STP) \cdot cm}{sec \cdot cm^2 \cdot cm\ Hg}$
Ideal separation factor $(\alpha^*_{O_2-N_2})$	3.4	3.4
Flow conditions	No mixing on either side of membrane	No mixing on either side of membrane
membrane area	11.5×10^6 ft^2	13.3×10^7 ft^2
Power requirements	0.34 kW-hr/100 ft^3 (product)	4 kW-hr/100 ft^3 (product)

plants has greatly increased in the last 15–20 yr. Second, the plant investment must also be taken into account to obtain a realistic evaluation of the permeation process. It now appears possible to construct permeation equipment for large-scale applications, such as described in the previous section, for as little as $1.00/ft^2 installed. Despite this low cost, the area of membrane required is so large that the permeation process cannot be considered as being of practical interest. Finally, it should be noted that the output of the permeation process under consideration is equivalent to that of a very small cryogenic plant for air separation. Since the product is oxygen-enriched air rather than high-purity oxygen, present-day economics would probably preclude the construction of such a small cryogenic plant. The enriched air would be produced by evaporation of liquid oxygen and dilution of the pure oxygen gas thus obtained with the appropriate amounts of air. Hence, any economic comparisons must be made at present on this basis.

Kammermeyer has suggested the use of silicone rubber membranes for the separation of oxygen from air, because their permeability to oxygen at ambient temperature is 30–60 times higher than that of ethyl cellulose membranes (depending on the origin of the former).[68, 69] Additionally, permeability can be increased substantially by operating at elevated temperatures, and silicone rubber can withstand much higher temperatures than other polymers. On the other hand, the selectivity of silicone rubber for oxygen is lower than that of ethyl cellulose: $\alpha^*_{O_2-N_2} = 2.1$–3.0 as compared to 3.4, respectively, at ambient temperature. Also, the selectivity decreases with increasing temperature. Thus operation at temperatures much above ambient may prove impractical.

A cursory evaluation indicates that a permeation process cannot compete at this time with conventional processes in the production of oxygen of any purity on a large scale, even when using 0.0001-in.-thick silicone rubber membranes. However, the permeation method may prove useful as a source of oxygen-enriched air for hospitals and other small-scale applications.

It is interesting to speculate whether improved membranes, which would exhibit both higher permeabilities and selectivities for oxygen, could be "tailored" at some future time. Table 3 summarizes the permeabilities to oxygen and nitrogen, and the corresponding ideal separation factors, for a large variety of membranes. It is seen that although the permeabilities vary by over five orders of magnitude, the separation factors vary only by a factor of

Table 3.   Permeability of Various Organic Membranes to Oxygen and Nitrogen at 30°C[66]

Membrane	Permeability Coefficient, $P_0 \times 10^{10} \left[\dfrac{cm^3(STP)\cdot cm}{sec\cdot cm^2\cdot cm\,Hg}\right]$		Ideal Separation Factor, $\alpha^*_{O_2-N_2}$
	O_2	N_2	
Saran	0.0053	0.00094	5.5
Mylar	0.022	0.005	4.4
Nylon 6	0.038	0.010	3.8
Kel-F	0.56	0.13	4.3
Pliofilm FM	0.54	0.14	4.0
Cellulose acetate P-912	0.78	0.28	2.8
Hycar OR 15	0.96	0.24	4.0
Butyl rubber	1.3	0.31	4.2
Vulcaprene	1.5	0.49	3.0
Methyl rubber	2.1	0.48	4.3
Polyethylene	3.0	1.3	2.3
Teflon FEP	6.2	2.0	3.1
Polystyrene	6.4	2.2	2.9
Ethyl cellulose	9.6	2.8	3.4
Polybutadiene	19	6.5	2.9
Natural rubber	30	12	2.7
Nitrile silicone rubber	81	21	3.4
Phenylene silicone rubber	140	40	3.5
Silicone rubber	600	260	2.3

about 2; moreover, the more permeable membranes exhibit lower selectivities. This behavior suggests that the chances of synthesizing more highly permeable membranes are considerably better than of finding membranes with greater selectivity toward oxygen.

B.   Recovery of Helium from Natural Gas

The recovery of helium from natural gas has been studied in perhaps greater detail than any other gas permeation process. Interest in this process arose from the observation[26, 70] that Teflon FEP, a copolymer of tetrafluoroethylene and hexafluoropropylene, exhibits both a high permeability and a high selectivity for helium as compared to the other components of natural gas (see Table 1).

In order to evaluate the economics of a permeation process for helium recovery, it was assumed that natural gas containing 0.45% helium, 17.06% nitrogen, 76.43% methane, and 6.06% higher hydrocarbons at a pipeline pressure of 900 psig was representative of the helium-bearing streams available for processing. A product gas containing at least 70% helium at a delivery pressure of 800 psig was required. To simplify the evaluation, it was further assumed that Teflon FEP was impermeable to hydrocarbons higher than methane. In tests with various gas mixtures it was found that this polymer is indeed less permeable to some of the higher hydrocarbons than to methane.

The calculations showed that three stages of permeation are necessary to obtain the desired helium concentration in the product stream. The optimum operating conditions are summarized in Table 4. Initial compression of natural gas is required to make up for unavoidable pressure drops through the plant feed lines, the first-stage permeator, and the return lines. All helium

Table 4.   Recovery of Helium from Natural Gas by Permeation Through Teflon FEP Membranes[26]

Optimum Operating Conditions for Large Plants
0.001-in.-thick membrane
Three permeation stages
Feed gas compressed to 1000 psig
80°C operating temperature
7.3-psia permeator back pressure
60% recovery of helium in natural gas
Interstage gas compressed to 950 psig
72% helium in product gas

losses (except for leakage losses) are in the first permeation stage, because the unpermeated gas from each succeeding stage is returned to the feed side of the preceding stage. Equipment must be provided for heating the feed gas prior to permeation and cooling the permeated gas before compression. Second- and third-stage feed streams can be heated to operating temperatures by the work of compression.

The effect of a number of design variables on the plant investment was calculated in some detail. The variables included operating temperature, feed gas compression, back pressure of permeated gas, efficiency of helium recovery from the feed, and plant size. It was found that the investment required is a minimum for large plants (i.e., those processing 500 million ft^3 of natural gas per day) when designed to recover about 60% of the helium in the feed gas. Very high recovery efficiencies are not practical with partially selective membranes such as Teflon FEP because very large permeators are required. In order to recover 60% of the total helium content, only

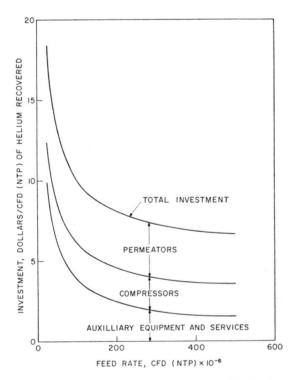

Figure 16. Plant investment as a function of feed rate (plant size) for a three-stage helium recovery process using Teflon FEP membranes.[26]

5.3% of the feed gas has to permeate through the first stage of the plant; the helium concentration in the permeate emerging from this stage is about 5%. The helium concentration is then raised to 25% in the second permeation stage and finally to 72% in the third stage.

Figure 16 shows the plant investment required for a three-stage Teflon FEP permeation plant for helium recovery as a function of plant size. The distribution of plant costs into three subtotals—permeators, compressors, and auxiliary equipment and services—is also shown. A number of relatively constant-cost items included with the auxiliary equipment service results in their becoming an increasingly larger fraction of the total investment as the plant feed rate is decreased.

The estimated total compressor power required for operation of the permeation plant is 0.215 kW-hr/ft^3 (NTP) of helium recovered. Since the power requirements are relatively low and can be satisfied with gas-driven equipment, power costs are only a small fraction of the total cost of recovering helium. Plant investment costs comprise 90–95% of the total cost.

The described process is not competitive with the most advanced cryogenic processes for the large-scale recovery of helium, mainly because of the large membrane area needed. However, a recently synthesized perfluoro membrane[71] exhibits a much higher permeability and selectivity towards helium than Teflon FEP, and may substantially improve the economics of the helium permeation process.

C. Carbon Dioxide Control in Breathing Atmospheres

One of the essential elements of life-support systems for sealed environments, such as space capsules, is a carbon dioxide controller capable of extracting approximately 2.25 lb of CO_2 per man-day from the atmosphere. Although a number of satisfactory methods of carbon dioxide removal are presently available for missions of various durations, efforts are continuing toward the development of more efficient and lighter weight systems. Membrane separation processes appear to hold promise of being able to achieve both of the above goals.

A preliminary calculation has been made to assess the weight and power requirement of a membrane process for carbon dioxide removal for a three-man mission. The calculation is based on the use of a hypothetical membrane with assumed permeability properties. For the sake of simplicity, an average permeability value was used for air. The results are summarized in Table 5 and are also shown schematically in Figure 17. With a feed gas consisting of air with 1% CO_2, a cascade of four permeation stages will produce an air-return stream containing 0.1% CO_2 and a waste stream with 90.5% CO_2. Only streams of the same composition are mixed at any stage. As usual for

Table 5.   A Multi-Stage Membrane Process for Carbon Dioxide Control[66]

Assumptions

Permeability coefficient, $P_0^{CO_2}$	$6 \times 10^{-8}\ \dfrac{cm^3(STP) \cdot cm}{sec \cdot cm^2 \cdot cm\ Hg}$
Ideal separation factor, $\alpha^*_{CO_2-Air}$	20
Membrane thickness	0.0005 in. $= 0.5\ mil!$
Feed composition	Air with $1\%\ CO_2$
Composition of return gas (product)	Air with $0.1\%\ CO_2$
Pressure ratio across membranes	20
Pressure of permeated gas	1 atm (abs.)

Operating Conditions

Stage	Feed	Permeate,	Permeate Purity (mole-% CO_2)	Membrane Area (ft²)	Permeator Weight (lb)	Power[a] (kW)
	(ft³(STP)/hr)					
1	293.0	58	0.055	1545.0	100	1.14
2	66.5	12	0.261	180.0	28	0.26
3	13.5	5	0.634	16.0	5	0.084
4	5.0	3	0.905	7.7	3	0.035
					136	1.519

Amount of power recovered in expansion engine = 0.730 kW
Net power consumption = 0.789 kW
Weight penalty for power = 300 lb/kW

[a] Adiabatic compression efficiencies assumed from 40 to 80%, depending on size of machine.

such a process, flow rates through successive stages are seen to decrease rapidly. Hence the weight and power requirements of the process are determined largely by the first stage. The return gas from the first stage, depleted in carbon dioxide, can be passed through an expansion engine to recover about 50–60% of the power used to compress the feed gas and the permeate streams.

The weight of the permeation equipment is seen to be only 136 lb. It is estimated that the weight penalty for power consumption plus the weight of heat exchangers amounts to about 270 lb. The total weight of the permeation system, about 406 lb, appears to be reasonably close to that of competing carbon dioxide removal systems.

The best membrane available at present for this process is silicone rubber, as suggested also by Kammermeyer.[68, 69] The permeability of silicone rubber to carbon dioxide is much larger than assumed in the above calculation, but its selectivity for this gas relative to air is much lower. It is possible that a more detailed calculation will show that a competitive membrane process

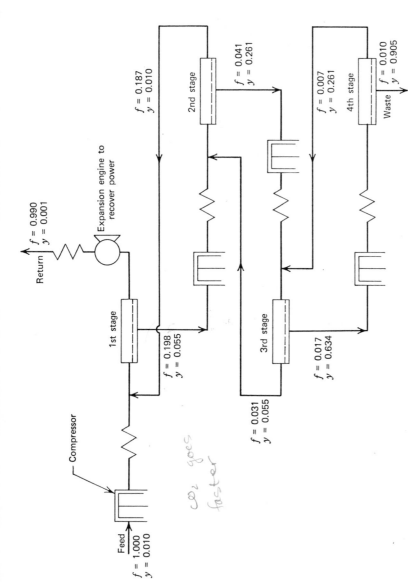

Figure 17. Four-stage permeation process for carbon dioxide control in breathing atmospheres.[66] f = Relative volume of the gas streams entering and leaving the various stages expressed as a fraction of the feed stream; y = mole-fraction of carbon dioxide in these streams.

331

for carbon dioxide control can be designed with silicone rubber or other existing membranes.

D. Removal of Rare Gas Fission Products

The removal of radioactive xenon and krypton from mixtures with other gases is of particular interest to the nuclear industry. The possibility of achieving this removal by selective permeation through silicone rubber membranes has been studied by the U.S. Atomic Energy Commission, and detailed experimental data and cost information are available on the subject.[72–75] The process could be used for the decontamination of the following gases: (*a*) the air within a reactor-containment building, after an accidental release of fission products; (*b*) the off-gas from a plant for processing spent reactor fuels; and (*c*) the gas that blankets certain types of nuclear reactors, such as the molten salt and the sodium-cooled breeder reactors, which continuously vent gaseous fission products. Figure 18 is a schematic diagram of a permeation plant for removing xenon and krypton from the argon cover gas of a 1000 MW(e) sodium-cooled fast reactor. Such a plant would continuously process a small stream of cover gas, and remove sufficient amounts of the rare gas fission products to allow more than 90% of the feed stream to be recycled to the reactor or be discharged to the atmosphere. The product from the top of the permeation plant, containing concentrated xenon and krypton, would be compressed to 2200 psig and stored in conventional gas cylinders. The performance, size, and plant cost

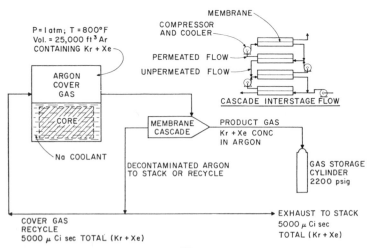

Figure 18.   Schematic diagram for removing xenon and krypton from the argon cover gas of a 1000 MW(e), sodium-cooled fast nuclear reactor. (From Ref. 75, courtesy of Union Carbide Corporation.)

Table 6. Performance and Cost of Several Cascades of Permselective Membranes for Separating Krypton and Xenon from Cover Gas of a Sodium-Cooled, 1000-MW(e) Reactor[75]

Gas Feed Rate to Cascade = 10 ft³(STP)/min
Membrane: 0.002-in.-thick Silicone Rubber

	Product Gas Rate (ft³(STP)/min.)		
	1	0.1	0.02
Cascade			
Number of enriching stages	9	15	20
Number of stripping stages	19	23	24
Membrane area (yd²)	4630	5020	5140
Cascade volume (ft³)	~9	~10	~11
Power requirement (kW)	126	136	140
Largest compressor (horsepower)	9.2	7.9	7.7
High-pressure side of membrane (psia)	150	150	150
Low-pressure side of membrane (psia)	0	0	0
Feed Gas			
Kr concentration (at. %)	0.004	0.004	0.004
Xe concentration (at. %)	0.136	0.136	0.136
Kr activity (Ci/ft³)	24.4	24.4	24.4
Xe activity (Ci/ft³)	542.6	542.6	542.6
Recycle (or vented) gas			
Kr concentration (at. %)	0.47×10^{-5}	0.49×10^{-5}	0.46×10^{-5}
Xe concentration (at. %)	0.53×10^{-9}	0.51×10^{-10}	0.25×10^{-10}
Kr + Xe activity (Ci/ft³)	0.029	0.030	0.028
Kr + Xe activity (μCi/sec)	4350	4950	4657
Product gas (to be stored)			
Kr concentration (at. %)	0.04	0.4	2
Xe concentration (at. %)	1.4	14	68
Ar concentration (at. %)	98.6	85.0	30
Kr activity (Ci/ft³)	244	2440	12,200
Xe activity (Ci/ft³)	5426	54,260	271,300
Concentration factor (product/feed)	10	100	500
Number storage cylinders per week[a]	50	5	1
Installed cost ($)	289,000	342,000	366,000

[a] Standard N_2 cylinders at a pressure of 2200 lb/in.²

have been calculated by Rainey, Blumkin, et al.[75] for three different product rates, and their results are reproduced in Table 6. These rates correspond to the recycling of 90, 99, and 99.8% of the feed, after reducing the radioactivity of the recycle gas to acceptable levels.

V. PERSPECTIVE

The main obstacle delaying the realization of competitive gas permeation processes, either from an economic standpoint or in terms of other criteria, is the relatively low permeability to gases of the polymeric membranes available at present. This low permeability has resulted in large membrane area requirements and correspondingly high capital investment costs. By comparison, the selectivity of the available membranes has generally proved to be adequate for separation purposes.

It has been mentioned recently[51] that the membrane area requirements of any permeation process could be reduced by one or more of the following methods: (a) development of more permeable membranes, (b) preparation of very thin membranes, and (c) optimization of operating conditions.

The development of more permeable membranes will have to rely to some extent on trial-and-error methods, because the effect of membrane composition and morphology on permeation mechanism is not sufficiently well understood. In this respect, a so-called tailoring of special membranes for specific gas permeation processes, in the true sense of the word, is not within the present state-of-the-art of polymer science. Nevertheless, progress is being made in this area, as witnessed by the synthesis of highly effective perfluoro membranes for helium recovery.[71] Various methods of preparing very thin separation membranes are also being investigated, and such membranes are being employed with striking success in the desalination of seawater by reverse osmosis. Finally, good progress is being made in all engineering aspects of gas permeation.

These favorable trends may lead one to conclude that the immediate prospects of gas permeation technology are encouraging. This writer feels obliged to inject a note of caution, which it is hoped will not dampen the enthusiasm of anyone whose imagination has been challenged by this interesting technology. The path leading to the development of a successful gas permeation process may prove to be tortuous and the rate of progress not as rapid as anticipated. Moreover, it should be remembered that the conventional separation processes against which permeation must compete are also being continuously improved. Therefore, the developer of any new process is shooting, so to speak, at a moving target. Nevertheless, the simplicity and versatility of gas permeation entirely justify the optimism of the investigators in this field. Future progress will depend, however, on the successful and purposeful blending of several disciplines, such as polymer science, physical chemistry, and separation technology and economics. In other words, the development of a competitive gas permeation process will require an interdisciplinary approach and, probably, the talents of a small team of investigators. In such a team, the predominant role must be played by the polymer scientist.

Finally, the main emphasis of this chapter has been on large-scale permeation processes, which are primarily of industrial interest, and the above conclusions were drawn in this context. It is highly probable that gas permeation will find other important applications, such as in biomedical devices, in the near future.

ACKNOWLEDGMENT

The author wishes to express his indebtedness to Dr. H. L. Frisch of State University of New York at Albany and to Dr. S. Blumkin of Oak Ridge Gaseous Diffusion Plant for their valuable comments on several sections of this chapter.

APPENDIX A

The *local or point separation factor* is defined by

$$\alpha = \frac{y^A/(1 - y^A)}{x^A/(1 - x^A)} \tag{A1}$$

The *local* permeation rates of the components A and B of a binary mixture across an element of membrane area dA are expressed, at steady-state, by Equations 61 and 62. The ratio of these equations is

$$\frac{y^A}{1 - y^A} = \left(\frac{P_0^A}{P_0^B}\right)\frac{x^A - (1/r)y^A}{(1 - x^A) - (1/r)(1 - y^A)} \tag{A2}$$

where $r(= p_h/p_l)$ is the ratio of total pressures on the two sides of the membrane $(p_h > p_l)$.

Equations A1 and A2 yield the expression

$$\alpha = \left(\frac{P_0^A}{P_0^B}\right)\left(\frac{(1 - x^A)/(1 - y^A)}{[(1 - x^A)/(1 - y^A)] + (1/r)[(P_0^A/P_0^B) - 1]}\right) \tag{A3}$$

where

$$\alpha^* = \frac{P_0^A}{P_0^B} \tag{A4}$$

is the ideal separation factor. Equations A1 and A3 yield the following expression for the local or point separation factor:

$$\alpha = \frac{(\alpha^* + 1)}{2} - \frac{(1/r)(\alpha^* - 1)}{2} - \frac{1}{2x^A} \pm \left\{ \left(\frac{\alpha^* - 1}{2} \right)^2 \right.$$

$$+ \frac{(\alpha^* - 1) - (1/r)[(\alpha^*)^2 - 1]}{2x^A} + \left[\frac{(1/r)(\alpha^* - 1) + 1}{2x^A} \right]^2 \right\}^{\frac{1}{2}} \quad \text{(A5)}$$

Only the positive root is used. Thus, α depends on α^*, the pressure ratio r, and the local composition x^A of the more permeable component on the high-pressure side of the membrane.

It is seen that

$$\alpha \to \alpha^* \quad \text{when} \quad r \to \infty$$

that is, the local separation factor reduces to the ideal separation factor when $p_l \to 0$.

APPENDIX B

In order to evaluate the membrane area for the "no mixing" case by means of Equation 70, it is necessary to express dL and y^A as a function of x^A.

Reference is made to Figure 10. A material balance around a differential volume in the high-pressure stream yields

$$\frac{dL}{L} = \frac{dx^A}{y^A - x^A} = \left[\frac{1 + \varepsilon x^A}{x^A \varepsilon (1 - x^A)} \right] dx^A \quad \text{(B1)}$$

where L is the total molar flow rate at any point on the high-pressure side of the stage and $\varepsilon (\equiv \alpha - 1)$ is the *local or point enrichment factor*. The above material balance equation is similar to that for Rayleigh (batch) distillation.

Under the assumption of constant enrichment (or separation) factor, Equation B1 can be integrated from the volume element to the stage outlet in order to yield an expression for L

$$L = L_{i(h)}(1 - \phi) \left[\left(\frac{x^A}{x_o^A} \right)^{1/\varepsilon} \left(\frac{1 - x_o^A}{1 - x^A} \right)^\sigma \right] \quad \text{(B2)}$$

where $\sigma = (\varepsilon + 1)/\varepsilon$. The product of Equations B1 and B2 gives dL as a function of x^A.

An expression for y^A as a function of x^A is obtained from the definition of the enrichment factor and Equation B1:

$$y^A = \frac{x^A(1 + \varepsilon)}{1 + \varepsilon x} \quad \text{(B3)}$$

REFERENCES

1. J. H. Mitchell, *J. Roy. Inst.* **2**, 101, 307 (1831).
2. T. Graham, *Phil. Mag.* **32**, 401 (1866).
3. P. Margis, D.R.P. 17,981, Class 12 (August 7, 1881).
4. M. Herzog, U.S. Pat. 307,041 (October 21, 1884).
5. S. W. Weller and W. A. Steiner, *J. Appl. Phys.* **21**, 279 (1950).
6. S. W. Weller and W. A. Steiner, *Chem. Eng. Prog.* **46**, 585 (1950).
7. D. W. Brubaker and K. Kammermeyer, *Ind. Eng. Chem.* **46**, 733 (1954).
8. K. Kammermeyer, *Ind. Eng. Chem.* **49**, 1685 (1957).
9. K. Kammermeyer, U.S. Pat. 2,966,235 (December 27, 1960).
10. G. J. van Amerongen, *Rubber Chem. Technol.* **37**, 1065 (1964).
11. C. E. Rogers, Chapter 6 in *Physics and Chemistry of the Organic Solid State*, D. Fox, M. M. Lobes, and A. Weissberger, Eds., Vol. II, Interscience, New York, 1965, pp. 510–627.
12. J. Crank and G. S. Park, Eds., *Diffusion in Polymers*, Academic, New York, 1968.
13. R. M. Barrer, *Diffusion in and Through Solids*, Cambridge University Press, 1941.
14. J. Crank, *The Mathematics of Diffusion*, Clarendon Press, Oxford, 1956.
15. W. Jost, *Diffusion in Solids, Liquids, Gases*, Academic, New York, 1952.
16. R. M. Barrer, *J. Phys. Chem.* **61**, 178 (1957).
17. A. Aitken and R. M. Barrer, *Trans. Faraday Soc.* **51**, 116 (1955).
18. C. E. Rogers, V. Stannett, and M. Szwarc, *J. Polymer Sci.* **45**, 61 (1960).
19. G. Gee and L. R. Treloar, *Trans. Faraday Soc.* **38**, 147 (1942).
20. J. Ferry, G. Gee, and L. R. G. Treloar, *Trans. Faraday Soc.* **41**, 340 (1945).
21. K. H. Meyer, E. Wolff, and C. Boissonnas, *Helv. Chim. Acta* **23**, 430 (1940).
22. P. Stamberger, *J. Chem. Soc.* **1929**, 2318.
23. J. Lens, *Rec. Trav. Chim.* **51**, 971 (1932).
24. R. M. Barrer, J. A. Barrie, and J. Slater, *J. Polymer Sci.* **23**, 315 (1957).
25. R. M. Barrer and J. A. Barrie, *J. Polymer Sci.* **23**, 331 (1957).
26. S. A. Stern, T. F. Sinclair, P. J. Gareis, N. P. Vahldieck, and P. H. Mohr, *Ind. Eng. Chem.* **57**, 49 (1965).
27. S. A. Stern, S.-M. Fang, and R. M. Jobbins, *J. Macromol. Sci. Phys.* (B) **5** (1), 41 (1971); see also *Polymer Abstr.* **10**, 1078 (1969).
28. R. Waack, N. H. Alex, H. L. Frisch, V. Stannett, and M. Szwarc, *Ind. Eng. Chem.* **47**, 2524 (1955).
29. S. A. Stern, J. T. Mullhaupt, and P. J. Gareis, *A.I.Ch.E. J.* **15**, 64 (1969).
30. S. A. Stern, S.-M. Fang, and H. L. Frisch, *J. Polymer Sci.*, A–2, **10**, 201 (1972).
31. J. A. Cutler, E. Kaplan, A. D. McLaren, and H. Mark, *Tappi* **34**, 404 (1951).
32. N. N. Li and R. B. Long, *A.I.Ch.E. J.* **15**, 73 (1969).
33. W. Heilman, V. Tammela, J. A. Meyer, V. Stannett, and M. Szwarc, *Ind. Eng. Chem.* **48**, 821 (1956).

34. A. W. Myers, J. A. Meyer, C. E. Rogers, V. Stannett, and M. Szwarc, *Tappi* **44**, 58 (1961).

35. R. M. Barrer and G. Skirrow, *J. Polymer Sci.* **3**, 564 (1948).

36. G. J. van Amerongen, *J. Polymer Sci.* **5**, 307 (1950).

37. S. A. Stern, P. J. Gareis, T. F. Sinclair, and P. H. Mohr, *J. Appl. Polymer Sci.* **7**, 2035 (1963).

38. P. Meares, *J. Am. Chem. Soc.* **76**, 3415 (1954).

39. P. Meares, *Trans. Faraday Soc.* **53**, 101 (1957).

40. I. Sobolev, J. A. Meyer, V. Stannett, and M. Szwarc, *Ind. Eng. Chem.* **49**, 441 (1957).

41. G. V. Casper and E. J. Henley, *J. Polymer Sci.* (B) **4**, 417 (1966).

42. L. M. Robeson and T. G. Smith, *J. Appl. Polymer Sci.* **12**, 2083 (1968).

43. H. L. Frisch, *J. Phys. Chem.* **60**, 1177 (1956).

44. A. S. Michaels, "Fundamentals of Membrane Permeation," *Proceedings of the Symposium on Membrane Processes for Industry, Southern Research Institute, Birmingham, Alabama, May 19–20, 1966*, p. 157.

45. H. E. Huckins and K. Kammermeyer, *Chem. Eng. Prog.* **49**, 180 (1953).

46. H. E. Huckins and K. Kammermeyer, *Chem. Eng. Prog.* **49**, 295 (1953).

47. R. W. Naylor and P. O. Backer, *A.I.Ch.E. J.* **1**, 95 (1955).

48. M. E. Breuer and K. Kammermeyer, *Sep. Sci.* **2**, 319 (1967).

49. J. Oishi, Y. Matsumura, K. Higashi, and C. Ike, *J. At. Energy Soc. Japan* **3**, 923 (1961); *U.S. At. Energy Comm. Rep. No. AEC-TR-5134*.

50. K. Kammermeyer, *Progress in Separation and Purification*, E. S. Perry, Ed., Vol. I, Interscience, New York, 1968, pp. 335–72.

51. S. A. Stern and W. P. Walawender, Jr., *Sep. Sci.* **4**, 129 (1969).

52. S. T. Hwang, *Sep. Sci.* **4**, 167 (1969).

53. K. Cohen, "The Theory of Isotope Separation as Applied to the Large Scale Production of U²³⁵," *National Nuclear Energy Series, Division III*, Vol. 1B, McGraw-Hill, New York, 1951.

54. S. T. Hwang and K. Kammermeyer, *Can. J. Chem. Eng.* **43**, 36 (1965).

55. S. Blumkin, "A Method for Calculating Cascade Gradients for Multicomponent Systems Involving Large Separation Factors," *Report K-OA-1559*, Union Carbide Corporation, Nuclear Division, Oak Ridge Gaseous Diffusion Plant, Oak Ridge, Tenn., January 1968.

56. J. O. Osburn and K. Kammermeyer, *Ind. Eng. Chem.* **46**, 739 (1954).

57. A. S. Michaels, U.S. Pat. 3,173,867 (September 28, 1962); Re. 26097 (October 11, 1966).

58. W. A. Steiner and S. W. Weller, U.S. Pat. 2,597,907 (May 27, 1952).

59. S. W. Weller, personal communication.

60. S. J. Megibow, U.S. Pat. 3,266,629 (August 16, 1966).

61. S. A. Stern, U.S. Pat. 3,332,216 (July 25, 1967).

62. H. W. McRobbie and L. M. Litz, "Reverse Osmosis Permeators—A Design Study," *Fifth Quarterly Report and Phase I Summary*, Union Carbide Corporation, Research Institute, Tarrytown, New York, October 1968, Office of Saline Water, U.S. Dept. of the Interior, Contract No. 14-01-0001-1293.

63. H. I. Mahon, E. A. McLain, W. E. Skeins, B. J. Green, and T. E. Davis, "Unusual Methods of Separation," *Chem. Eng. Prog. Symp. Ser.* **65**, 48 (1969).

64. W. T. Robinson and R. J. Mattson, *J. Water Pollut. Control Fed.* **40**, Part 1, 439 (1968).

65. R. J. Mattson and V. J. Tomsic, *Chem. Eng. Prog.* **65**, 62 (1969).

66. S. A. Stern, "Industrial Applications of Membrane Processes: The Separation of Gas Mixtures," *Proceedings of the Symposium on Membrane Processes for Industry, Southern Research Institute, Birmingham, Alabama, May 19–20, 1966*, p. 196.

67. S. W. Weller and W. A. Steiner, U.S. Pat. 2,540,151 (February 6, 1951).

68. K. Kammermeyer, *Ind. Eng. Chem.* **49**, 1685 (1957).

69. K. Kammermeyer, U.S. Pat. 2,966,235 (December 27, 1960).

70. S. A. Stern, P. H. Mohr, and P. J. Gareis, U.S. Pat. 3,246,499 (April 19, 1966).

71. E. L. Niedzielski and R. E. Putnam, U.S. Pat. 3,307,330 (March 7, 1967).

72. R. H. Rainey, W. L. Carter, S. Blumkin, and D. E. Fain, "Treatment of Airborne Radioactive Wastes," *Intern. At. Energy Agency, Tech. Rep. Ser., Vienna, 1968*, p. 323.

73. S. Blumkin, "Separation of Krypton from Air by Means of a Permselective Membrane," *Report K-OA-1552*, Union Carbide Corporation, Nuclear Division, Oak Ridge Gaseous Diffusion Plant, Oak Ridge, Tenn., November 1967.

74. S. Blumkin, "Permselective Membrane Cascade for the Separation of Xenon and Krypton from Argon," *Report K-OA-1622*, Union Carbide Corporation, Nuclear Division, Oak Ridge Gaseous Diffusion Plant, Oak Ridge, Tenn., July 1968.

75. R. H. Rainey, W. L. Carter, and S. Blumkin, "Evaluation of the Use of Permselective Membranes in the Nuclear Industry for Removing Radioactive Xenon and Krypton from Various Off-Gas Streams," *Report ORNL-4522*, Oak Ridge National Laboratory, Oak Ridge, Tenn., April 1971.

Index